T0255028

LONDON MATHEMATICAL SOCIETY LECTURE NOTE SERIES

Managing Editor: Professor M. Reid, Mathematics Institute,
University of Warwick, Coventry CV4 7AL, United Kingdom

The titles below are available from booksellers, or from Cambridge University Press at
http://www.cambridge.org/mathematics

London Mathematical Society Lecture Note Series: 421

O-Minimality and Diophantine Geometry

Edited by

G. O. JONES
University of Manchester

A. J. WILKIE
University of Manchester

CAMBRIDGE
UNIVERSITY PRESS

University Printing House, Cambridge CB2 8BS, United Kingdom

Cambridge University Press is part of the University of Cambridge.

It furthers the University's mission by disseminating knowledge in the pursuit of education, learning and research at the highest international levels of excellence.

www.cambridge.org
Information on this title: www.cambridge.org/9781107462496

© Cambridge University Press 2015

This publication is in copyright. Subject to statutory exception and to the provisions of relevant collective licensing agreements, no reproduction of any part may take place without the written permission of Cambridge University Press.

First published 2015

A catalogue record for this publication is available from the British Library

Library of Congress Cataloguing in Publication data
Wilkie, A. J. (Alec J.)
O-minimality and diophantine geometry / G.O. Jones, University of Manchester, A.J. Wilkie, University of Manchester.
pages   cm. – (London Mathematical Society lecture note series ; 421)
ISBN 978-1-107-46249-6 (pbk.)
1. Arithmetical algebraic geometry.   2. Model theory.   3. Geometry, Analytic.
I. Jones, G. O. (Gareth Owen).   II. Title.
QA242.5.W55 2015
516.3′5–dc23   2014045023

ISBN 978-1-107-46249-6 Hardback

Cambridge University Press has no responsibility for the persistence or accuracy of URLs for external or third-party internet websites referred to in this publication, and does not guarantee that any content on such websites is, or will remain, accurate or appropriate.

Contents

Preface

In July 2013 an LMS-EPSRC Short Instructional Course on 'O-minimality and diophantine geometry' was held in the School of Mathematics at the University of Manchester. This volume consists of lecture notes from the courses together with several other surveys. The motivation behind the short course was to introduce participants to some of the ideas behind Pila's recent proof of the André-Oort conjecture for products of modular curves. The underlying ideas are similar to an earlier proof by Pila and Zannier of the Manin-Mumford conjecture (which has in fact long been a theorem, originally due to Raynaud) and combining the results of the various contributions here leads to a proof of this conjecture in certain cases. The basic strategy has three main ingredients: the Pila-Wilkie theorem, bounds on Galois orbits, and functional transcendence results. Each of the topics was the focus of a course. Wilkie discussed o-minimality and the Pila-Wilkie theorem without assuming any background in mathematical logic. (The argument given here is, in fact, slightly different from that given in the original paper, at least in the one-dimensional case.) Habegger's course focused on the Galois bounds and on the completion of the proof (of certain cases of Manin-Mumford) from the various ingredients. And Pila's lectures covered functional transcendence, also touching on various recent related work by Zilber. We have also included some further lecture notes by Wilkie containing a proof of the o-minimality of the expansion of the real field by restricted analytic functions, which is sufficient for the application of Pila-Wilkie to Manin-Mumford. At the short course there were also three guest lectures. Yafaev spoke on very recent breakthroughs on the functional transcendence side in the setting of general Shimura varieties. Masser spoke on some other results ('relative Manin-Mumford') that can be obtained by a similar strategy. Jones discussed improvements to the Pila-Wilkie theorem. Unfortunately, Yafaev was unable to contribute to this volume. During the week of the course, tutorials were given by Daw and Orr. For this volume,

Orr has written a survey of abelian varieties which contains a proof of the functional transcendence result necessary for the application in Habegger's course. Daw has contributed an introduction to Shimura varieties which we hope will prove valuable to those who wish to go on to study the general André-Oort conjecture. Finally, we are pleased to include a paper by Tsimerman in which he gives a proof of Ax's theorem on the functional case of Schanuel's conjecture via o-minimality.

We would like to thank the London Mathematical Society and the Engineering and Sciences Research Council for funding the course, and the School of Mathematics at the University of Manchester for hosting the meeting. And we are grateful to the speakers and tutors at the meeting and to the contributors to this volume.

1

The Manin-Mumford Conjecture, an elliptic Curve, its Torsion Points & their Galois Orbits

P. Habegger

Abstract

This is an extended write-up of my five hour lecture course in July 2013 on applications of o-minimality to problems in Diophantine Geometry. The course covered arithmetic properties of torsion points on elliptic curves and how they combine with the Pila-Wilkie Point Counting Theorem and the Ax-Lindemann-Weierstrass Theorem to prove a special case of the Manin-Mumford Conjecture.

1 Overview

These notes are a write-up of my lecture course titled *Diophantine Applications* which was part of the *LMS-EPSRC Short Instructional Course – O-Minimality and Diophantine Geometry* in Manchester, July 2013. The purpose of the short course was to present recent developments involving the interaction of methods from Model Theory with problems in Number Theory, most notably the André-Oort and Manin-Mumford Conjectures, to an audience of students in Model Theory and Number Theory.

At the heart of this connection is the powerful Pila-Wilkie Counting Theorem [26]. It gives upper bounds for the number of rational points on sets which are definable in an o-minimal structure.

The Manin-Mumford Conjecture concerns the distribution of points of finite order on an abelian variety with respect to the Zariski topology. We give a rather general version of this conjecture, later on we often work in the situation

O-Minimality and Diophantine Geometry, ed. G. O. Jones and A. J. Wilkie. Published by Cambridge University Press. © Cambridge University Press 2015.

where the base field is $\overline{\mathbf{Q}}$, the algebraic closure of $\mathbf{Q}$ inside the field of complex numbers $\mathbf{C}$. But soon we concentrate on the power of an elliptic curve.

Theorem 1.1 (Raynaud [32]) *Let A be an abelian variety defined over* $\mathbf{C}$. *Let* $\mathcal{X}$ *be an irreducible closed subvariety of A. We write*

$$A_{tors} = \{P \in A(\mathbf{C}); \ P \text{ has finite order}\}$$

for the group of all torsion points of A. Then $\mathcal{X}(\mathbf{C}) \cap A_{tors}$ *is Zariski dense in* $\mathcal{X}$ *if and only if* $\mathcal{X}$ *is an irreducible component of an algebraic subgroup of A.*

Any algebraic subgroup of A is a finite union of translates of an irreducible algebraic subgroup by points of finite order. The theory of abelian varieties guarantees that the torsion points lie Zariski dense on any algebraic subgroup. Showing that torsion points do not lie Zariski dense on a subvariety that is not a component of an algebraic subgroup is the difficult part of the Manin-Mumford Conjecture.

Prior to Raynaud's proof he was able to handle the case of a curve $\mathcal{X}$ [31]. Earlier partial results are due to Bogomolov [3, 4].

Lang [17] was interested in the analogous problem with A replaced by $(\mathbf{C}^{\times})^n$, where $R^{\times}$ denotes the unit group of any ring R. Here the points of finite order are those whose coordinates are roots of unity. In his paper, Lang presents proofs of the Manin-Mumford Conjecture for $(\mathbf{C}^{\times})^2$ attributed to Ihara, Serre, and Tate independently. In a paper published in the same year, Mann [20] treated hypersurfaces in any power of $\mathbf{C}^{\times}$.

Later, Hindry [12] proved the generalization to all semi-abelian varieties defined over $\mathbf{C}$.

In the mean time new proofs of variants of the Manin-Mumford Conjecture using various techniques have appeared in the literature: Hrushovski [16] used the Model Theory of difference fields, Pink-Rössler [28, 29] used classical Algebraic Geometry, and Ratazzi-Ullmo [30] used equidistribution.

Based on a strategy due to Zannier, he himself and Pila [27] used the aforementioned counting theorem and lower bounds for the *Galois orbit* of torsion points to give yet another proof of the Manin-Mumford Conjecture for abelian varieties if the base field is $\overline{\mathbf{Q}}$. This general technique had broad implications for *open* problems in diophantine geometry such as the André-Oort Conjecture [25].

One aim of the short course was to present the ingredients required to prove the Manin-Mumford Conjecture for an algebraic curve inside a product of elliptic curves using the approach laid out in [27]:

- The Pila-Wilkie Counting Theorem.
- An o-minimal approach to the Ax-Lindemann-Weierstrass Theorem which is a special case of Schanuel's Conjecture in functional setting.
- Bounding from below the size of a Galois orbit of point of finite order on an elliptic curve.

This lecture concerns the last part. In Section 2.1 we will give a brief introduction to the relevant parts of the theory of elliptic curves. In Section 2.2 we will see how to attach a function, definable in some o-minimal structure, to the uniformizing map coming from the Weierstrass function. The uniformizing map establishes the link between points of finite order on the elliptic curve and rational points.

On the arithmetic side we will investigate the Galois orbit of a torsion point in Section 3.

Suppose we are presented with an elliptic curve E defined by an equation with coefficients in a number field K and a K-rational point T on E of finite order n. In order to get the method running we require a lower bound

$$[K(T) : K] \geq cn^\delta \tag{1.1}$$

where $c > 0$ and $\delta > 0$ are constants that are allowed to depend on E but not on T. The left-hand side is precisely the size of the Galois orbit

$$\left\{ \sigma(T); \ \sigma \in \mathrm{Gal}(\overline{K}/K) \right\}.$$

The crucial feature of (1.1) is the *polynomial* dependency in n. It is needed to compete with the upper bound coming from the Pila-Wilkie Theorem, as we will see in Section 4. In the multiplicative setting, the lower bound analogous to (1.1) follows from the most basic facts on cyclotomic fields and Euler's totient function. For elliptic curves, one can quite easily prove a sub-polynomial lower bound for $[K(T) : K]$. But breaking the polynomial barrier involves more care than in the multiplicative case.

In Section 4 we combine our efforts and rely also on results presented in Jonathan Pila's, Martin Orr's, and Alex Wilkie's notes in this volume to give a proof based on [27] of the Manin-Mumford Conjecture for curves in the power E^g.

Theorem 1.2 *Let E be an elliptic curve defined over a number field K contained in $\mathbf{C}$ and suppose $\mathcal{X} \subseteq E^g$ is an irreducible algebraic curve also defined over K. Then $\mathcal{X}(\mathbf{C}) \cap E^g_{tors}$ is infinite if and only if $\mathcal{X}$ is an irreducible component of an algebraic subgroup of E^g.*

In the appendix we give a proof of a special case of a theorem of Elkies for local height functions on elliptic curves. This inequality leads to a polynomial lower bound for the Galois orbit of a torsion point on an elliptic curve.

Most of the material presented in these lecture notes is classical. No emphasis was made to formulate things in their proper generality; the presentation was chosen to give a pragmatic introduction to the tools required to prove Theorem 1.2. Two excellent starting points for a more detailed overview of the theory of elliptic curves from the arithmetic point of view (and beyond) are books of Silverman [36] and Cassels [7]. For the theory of heights, which also plays an important role in this course, we refer to books of Bombieri and Gubler [5] or Hindry and Silverman [14].

2 Elliptic Curves

2.1 The Group Law and Points of Finite Order

Let K be a subfield of $\mathbf{C}$. An elliptic curve E defined over K is a smooth, projective curve of genus 1 with a prescribed K-rational point P_0. In this situation E can be represented by a Weierstrass equation

$$y^2 = 4x^3 - g_2 x - g_3 \tag{1.2}$$

where $g_2, g_3 \in K$ satisfy $g_2^3 - 27g_3^2 \neq 0$. This condition guarantees that we obtain a smooth curve.

Certainly, (1.2) defines an affine curve, whereas the elliptic curve E is projective by definition. It is silently understood that E is isomorphic to the projective curve in $\mathbf{P}^2$ cut out by the homogenized equation

$$y^2 z = 4x^3 - g_2 xz^2 - g_3 z^3.$$

The K-rational point P_0 then corresponds to $[0:1:0]$. It is the only point missing from the affine curve defined by (1.2).

The Weierstrass equation is by no means uniquely determined by (E, P_0). Indeed, any $u \in K^\times$ can be used to make a change of coordinates

$$(x', y') = (u^2 x, u^3 y)$$

and obtain a new Weierstrass equation

$$y'^2 = 4x'^3 - g_2' x' - g_3' \quad \text{with} \quad g_2' = u^4 g_2 \quad \text{and} \quad g_3' = u^6 g_3$$

for E.

One of the most basic, but important, facts is that the points of E carry the structure of an abelian group with neutral element P_0. There are several ways to

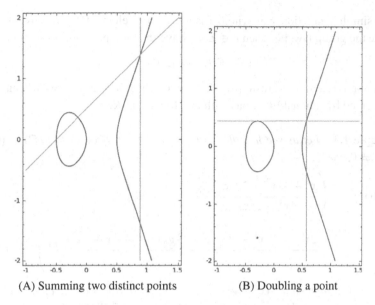

(A) Summing two distinct points   (B) Doubling a point

Figure 1.1 The chord and tangent construction for $y^2 = 4x^3 - x$

define the group law. Possibly the most straightforward method is via the well known chord and tangent construction which has a very geometric flavor. The procedure is illustrated in Figure 1.1. Given two *distinct* points on the affine portion of E we connect them by a line. This line intersects the elliptic curve in a third point which, after changing the sign of the y-coordinate, is the sum of the original points, cf. 1.1A. If the original points differ in the y-coordinate only, then their sum is the neutral element. We extend this to a binary operation on all points of E by treating P_0 as the neutral element in a group law. Adding a point on the affine part to itself requires some more care; now we intersect the tangent of E at this point to get a third point. The duplicate of the original point is obtained by again flipping the sign of the y-coordinate, cf. 1.1B. Of course, we define the duplication of P_0 to be again P_0.

The binary operation described above is given by rational functions with coefficients in K. With the exception of the law of associativity it is easy to see that this binary operation defines an abelian group law on the K-rational points of E with neutral element P_0; see Chapter III.2 [36] for explicit formulas. At least in principle it should be possible to verify associativity by an elaborate computation. On the other hand, there are approaches using the Riemann-Roch Theorem (Proposition 3.4 in Chapter III [36]) and a geometric one in Chapter 7 [7].

To simplify notation we write 0 for the neutral element P_0 and use $+$ to denote the group law. For each $n \in \mathbf{Z}$ we have the multiplication-by-n map

$$[n] : E(K) \to E(K).$$

It is a non-constant morphism $[n] : E \to E$ of algebraic curves. So $[n]$ can be represented by rational functions with coefficients in K.

Example 1.3 *Let us see how things look for $n = 2$. If $(x, y) \in E(\mathbf{C}) \smallsetminus \{0\}$ and $y \neq 0$, then*

$$[2](x, y) = \left(\frac{x^4 + \frac{g_2}{2}x^2 + 2g_3x + \frac{g_2^2}{16}}{4x^3 - g_2x - g_3}, \right.$$

$$\left. \frac{2x^6 - \frac{5g_2}{2}x^4 - 10g_3x^3 - \frac{5}{8}g_2^2x^2 - \frac{g_2g_3}{2}x + \frac{g_2^3}{32} - g_3^2}{y^3} \right), \quad (1.3)$$

which is well defined as $4x^3 - g_2x - g_3 = y^2 \neq 0$. Roots of the cubic $4x^3 - g_2x - g_3$ are precisely the x-coordinates of the points in $E(\mathbf{C})$ of order 2.

Definition 1.4 The group of torsion points of E is

$$E_{\text{tors}} = \{T \in E(\mathbf{C}); \text{ there exists an integer } n \geq 1 \text{ with } [n](T) = 0\}.$$

If $n \in \mathbf{N} = \{1, 2, 3, \ldots\}$ we write

$$E[n] = \{T \in E_{\text{tors}}; \ [n](T) = 0\}$$

for the group of points of finite order dividing n.

The structure of E_{tors} and $E[n]$ as abelian groups is well known. We will uncover both in the next section using the Weierstrass function.

Torsion points of E are algebraic over K, i.e.

$$E_{\text{tors}} = \{T \in E(\overline{K}); \ T \in E_{\text{tors}}\}$$

where $\overline{K}$ is the algebraic closure of K in $\mathbf{C}$. We reproduce this well known proof here. It involves the action of $\text{Aut}(\mathbf{C}/K)$, the field automorphisms of $\mathbf{C}$ that fix elements of K, a central concept for our arguments later on. If $P \in E(\mathbf{C})$, then

$$P \mapsto \sigma(P)$$

defines an automorphism of $E(\mathbf{C})$ as an abelian group; here σ acts on the coordinates of P if $P \neq 0$ and $\sigma(0) = 0$. Recall that $[n]$ is represented by rational functions with coefficients in K and so

$$\sigma([n](P)) = [n](\sigma(P)).$$

Say T has finite order $n \geq 1$, then so does $\sigma(T)$. In particular, the orbit

$$\{\sigma(T); \ \sigma \in \mathrm{Aut}(\mathbf{C}/K)\} \qquad (1.4)$$

is contained in a fiber of $[n]$. As $[n]$ is a non-constant morphism, all these fibers are finite and in particular $E[n]$ is finite. Thus (1.4) is finite. In particular, T cannot have a coordinate that is transcendental over K. This yields $T \in E(\overline{K})$, as desired. So already the Galois group $\mathrm{Gal}(\overline{K}/K)$ acts on $E[n]$ and E_{tors}.

It is not difficult to adapt the argument above to use only the duplication morphism [2], which we described explicitly in Example 1.3, and its iterates $[2^n]$. Indeed, by the Pigeonhole Principle $T \in E(\mathbf{C})$ is torsion if and only if $[2^n](T) = [2^m](T)$ for integers $0 \leq n < m$.

2.2 Uniformizing the complex points E

Here we describe an elliptic curve E from the analytic point of view. In the end it is the interplay between the algebraic and the analytic world that makes the strategy described in Section 1 feasible. Our goal is to attach a definable function to the inverse of the uniformizing map determined by a Weierstrass equation.

Suppose E is presented by the Weierstrass equation (1.2) with $g_2, g_3 \in \mathbf{C}$. There is a unique discrete, rank 2 subgroup $\Omega \subseteq \mathbf{C}$, called the periods of E, with the following properties.

- The series

$$\wp(z) = \frac{1}{z^2} + \sum_{\omega \in \Omega \smallsetminus \{0\}} \frac{1}{(z-\omega)^2} - \frac{1}{\omega^2}$$

determines a meromorphic, Ω-periodic function with poles of order 2 at points of Ω and no poles in $\mathbf{C} \smallsetminus \Omega$. It is called the Weierstrass function attached to (1.2). Moreover, this function satisfies the differential equation

$$\wp'^2 = 4\wp^3 - g_2\wp - g_3$$

and induces a surjective, analytic homomorphism of groups $u : \mathbf{C} \to E(\mathbf{C})$ defined by

$$u : z \mapsto \begin{cases} [\wp(z) : \wp'(z) : 1] & : \text{if } z \in \mathbf{C} \smallsetminus \Omega, \\ [0 : 1 : 0] & : \text{if } z \in \Omega \end{cases}$$

with kernel Ω.
- The coefficients in (1.2) are related to the periods by

$$g_2 = 60 \sum_{\omega \in \Omega \smallsetminus \{0\}} \frac{1}{\omega^4} \quad \text{and} \quad g_3 = 140 \sum_{\omega \in \Omega \smallsetminus \{0\}} \frac{1}{\omega^6}.$$

The existence of Ω follows from the theory of modular functions, see for example Proposition 5, Chapter VII [34]. Weierstrass functions are studied in Chapter VI [36].

Let us fix a **Z**-basis (ω_1, ω_2) of the periods Ω. We can think of $u(v_1\omega_1 + v_2\omega_2)$ as a function in the real coordinates v_1 and v_2. In these coordinates, the period lattice Ω becomes $\mathbf{Z}^2$. So u induces a surjective group homomorphism $\mathbf{R}^2 \to E(\mathbf{C})$ with kernel $\mathbf{Z}^2$. Apart from the integral points $\mathbf{Z}^2$ this homomorphism takes values in the affine part of E, i.e. the solutions in $\mathbf{C}^2$ of (1.2). The preimage of E_{tors} is precisely $\mathbf{Q}^2$ and we get isomorphisms of groups

$$E_{\text{tors}} \cong (\mathbf{Q}/\mathbf{Z})^2 \quad \text{and} \quad E[n] \cong (\mathbf{Z}/n\mathbf{Z})^2 \tag{1.5}$$

for all $n \in \mathbf{N}$.

Our goal is to uniformize the affine part of E by a function that is definable in the o-minimal structure $\mathbf{R}_{\text{an}}$, the o-minimal structure generated by restricted real analytic functions. A reasonable candidate is

$$[-1/2, 1/2]^2 \smallsetminus \{(0,0)\} \to \mathbf{C}^2 = \mathbf{R}^4 \tag{1.6}$$
$$(v_1, v_2) \mapsto (\wp(v_1\omega_1 + v_2\omega_2), \wp'(v_1\omega_1 + v_2\omega_2))$$

where we identify the target $\mathbf{C}^2$ with $\mathbf{R}^4$ by taking real and imaginary parts on both factors of $\mathbf{C}^2$.

However, we must take some care, as (1.6) does not extend to a real analytic function on an open neighborhood of the compact set $[-1/2, 1/2]^2$ due to the pole at $(0,0)$ This issue is not too severe. We must merely remind ourselves that the Weierstrass function $\wp$ has a double pole at $z = 0$ and hence $\wp'$ has a triple pole there. In a sufficiently small neighborhood $(-\epsilon, \epsilon)^2$ of $(0,0)$ the mapping $(v_1, v_2) \mapsto \wp'(v_1\omega_1 + v_2\omega_2)$ does not vanish and

$$(v_1, v_2) \mapsto \left(\frac{\wp(v_1\omega_1 + v_2\omega_2)}{\wp'(v_1\omega_1 + v_2\omega_2)}, \frac{1}{\wp'(v_1\omega_1 + v_2\omega_2)} \right) \in \mathbf{C}^2 \tag{1.7}$$

is real analytic on $(-\epsilon, \epsilon)^2$ if we send $(0,0)$ to 0. The mapping (1.7) composed with

$$(z, w) \mapsto \begin{cases} (zw^{-1}, w^{-1}) & : \text{if } w \neq 0, \\ (0, 0) & : \text{otherwise} \end{cases} \tag{1.8}$$

coincides outside of $(0,0)$ with (1.6). Now (1.8) is semi-algebraic and therefore definable in $\mathbf{R}_{\text{an}}$. As the composite of two definable functions is again definable we find that (1.6) is definable in $\mathbf{R}_{\text{an}}$ when restricted to $(-\epsilon, \epsilon)^2 \smallsetminus \{(0,0)\}$.

Now that we have handled the singularity at the origin, definability of (1.6) in $\mathbf{R}_{\text{an}}$ is straightforward. Indeed, its restriction to the compact set

$[-1/2, 1/2]^2 \setminus (-\epsilon, \epsilon)^2$ clearly extends to a real analytic map on some larger open set; take for example $[-3/4, 3/4]^2 \setminus [-\epsilon/2, \epsilon/2]^2$.

For technical reasons, i.e. to achieve injectivity, it is convenient to restrict (1.6) further to $(-1/2, 1/2]^2 \setminus \{(0, 0)\}$. This does not affect the definability property we just proved. We thus obtain a bijection

$$(-1/2, 1/2]^2 \setminus \{(0, 0)\} \to \{(x, y) \in \mathbf{C}^2; \ y^2 = 4x^3 - g_2 x - g_3\}$$
$$= E(\mathbf{C}) \setminus \{0\}$$

which is definable in $\mathbf{R}_{an}$. We will work with the inverse map

$$\xi : E(\mathbf{C}) \setminus \{0\} \to (-1/2, 1/2]^2, \tag{1.9}$$

which is also definable in $\mathbf{R}_{an}$.

As we started out with a group homomorphism we find $\xi(E_{tors} \setminus \{0\}) \subseteq \mathbf{Q}^2$. More precisely, if $n \in \mathbf{N}$ and $T \in E_{tors} \setminus \{0\}$ has order dividing n, then $\xi(T) \in \frac{1}{n}\mathbf{Z}^2$.

This concludes our discussion on definability properties of a single Weierstrass function. Peterzil and Starchenko [24] studied the definability question for a family of Weierstrass functions. In this generality one needs the larger o-minimal structure $\mathbf{R}_{an, exp}$ generated by $\mathbf{R}_{an}$ and the exponential function on the reals.

3 Galois Orbits of Torsion Points and Heights

3.1 The Arithmetic of Torsion Points

In this section we discuss Galois theoretic properties of torsion points on an elliptic curve E. As usual, we assume that E is presented by a Weierstrass equation (1.2) with coefficients g_2, g_3 in a field $K \subseteq \mathbf{C}$. Now we will assume in addition that K is a number field.

Roughly speaking, torsion points of E share many arithmetic properties with the roots of unity

$$\mu = \{\zeta \in \mathbf{C}^\times; \ \text{there is } n \in \mathbf{N} \text{ with } \zeta^n = 1\}.$$

Let us also write

$$\mu[n] = \{\zeta \in \mu; \ \zeta^n = 1\}$$

for all $n \in \mathbf{N}$.

In the table below we list some similarities between roots of unity and the torsion points of the elliptic curve E. We retain notation from Section 2.2 and use φ to denote Euler's totient function.

	Roots of unity	Torsion on E
Analytic description	$e^{2\pi\sqrt{-1}\nu}$ with $\nu \in \mathbf{Q}$	$u(\nu_1\omega_1 + \nu_2\omega)$ with $\nu_{1,2} \in \mathbf{Q}$
Group structure	$\mu \cong \mathbf{Q}/\mathbf{Z}$ $\mu[n] \cong \mathbf{Z}/n\mathbf{Z}$	$E_{\text{tors}} \cong (\mathbf{Q}/\mathbf{Z})^2$ $E[n] \cong (\mathbf{Z}/n\mathbf{Z})^2$
Order vs. degree	If ord$(\zeta) = n$, then $[\mathbf{Q}(\zeta) : \mathbf{Q}] = \varphi(n) \geq c\frac{n}{\log\log 3n}$	If ord$(T) = n$, then $[K(T) : K] \leq n^2$
Field properties	$\mathbf{Q}(\mu[n])/\mathbf{Q}$ Galois with group $(\mathbf{Z}/n\mathbf{Z})^\times = \text{GL}_1(\mathbf{Z}/n\mathbf{Z})$	$K(E[n])/K$ Galois with group isomorphic to a subgroup of $\text{GL}_2(\mathbf{Z}/n\mathbf{Z})$

We now justify the entries in this table. The middle column is classical algebra. The inequality

$$\varphi(n) \geq c\frac{n}{\log\log 3n} \qquad (1.10)$$

holds for all $n \geq 1$ where $c > 0$ is an absolute constant by Theorem 328 [10]. Any ζ of order n generates $\mu[n]$. So if $\sigma \in \text{Gal}(\mathbf{Q}(\zeta)/\mathbf{Q})$ then

$$\sigma(\zeta) = \zeta^a$$

for some exponent $a \in \mathbf{Z}$ that is uniquely determined modulo n. As σ is invertible, a must be coprime to n. We obtain a representation

$$\chi_n : \text{Gal}(\mathbf{Q}(\mu[n])/\mathbf{Q}) \to (\mathbf{Z}/n\mathbf{Z})^\times = \text{GL}_1(\mathbf{Z}/n\mathbf{Z})$$

determined by

$$\sigma(\zeta) = \zeta^{\chi_n(\sigma)}.$$

The representation χ_n is independent of the choice of the generator ζ of $\mu[n]$.

The analytic and group theoretic properties of the right column were discussed in Section 2.2 around (1.5). The upper bound for $[K(T) : K]$ follows as $\text{Gal}(\overline{K}/K)$ acts on the points of finite order n, of which there are at most n^2. However, if we return for a moment to the larger picture we require *lower bounds* for the Galois orbit of a torsion point to compete with the upper bound from the Pila-Wilkie Theorem. The easily obtainable upper bound in the table is in the wrong direction.

The action of the Galois group described towards the end of Section 2.1 guarantees that $K(E[n])/K$ is a Galois extension and that there is a natural representation of groups

$$\mathrm{Gal}(K(E[n])/K) \twoheadrightarrow \mathrm{Aut}\, E[n]$$

which target the group automorphisms of $E[n]$. Because $E[n] \cong (\mathbf{Z}/n\mathbf{Z})^2$ by (1.5) we may fix a $\mathbf{Z}/n\mathbf{Z}$-basis of $E[n]$. This leads to an injective group homomorphism

$$\rho_n : \mathrm{Gal}(K(E[n])/K) \to \mathrm{GL}_2(\mathbf{Z}/n\mathbf{Z})$$

and justifies the lower right entry in our table above. In contrast to the roots of unity, this representation may depend on the choice of basis. This ambiguity is harmless if one is only interested in the cardinality of the image $\mathrm{im}\, \rho_n$.

Example 1.5 *We cannot expect ρ_n to be a group isomorphism. Indeed, say $K = \mathbf{Q}$ and suppose E is given by*

$$y^2 = 4x^3 - 4x.$$

The points in $E[2] = \{0, (0,0), (\pm 1, 0)\}$ are defined over $\mathbf{Q}$. So

$$\rho_2 : \mathrm{Gal}(\mathbf{Q}(E[2])/\mathbf{Q}) \to \mathrm{GL}_2(\mathbf{Z}/2\mathbf{Z})$$

has trivial image. This affects the image of ρ_{2n} for all $n \in \mathbf{N}$. Indeed, if $T \in E[2n]$ then $[n](T) \in E(\mathbf{Q})$. So the image of ρ_{2n} is contained in the kernel of

$$\mathrm{GL}_2(\mathbf{Z}/2n\mathbf{Z}) \to \mathrm{GL}_2(\mathbf{Z}/2\mathbf{Z}).$$

We observe that this kernel has bounded index in $\mathrm{GL}_2(\mathbf{Z}/2n\mathbf{Z})$ as n varies.

For this elliptic curve there is a more severe restriction on the image of ρ_n. Over the number field $K = \mathbf{Q}(\sqrt{-1})$ the mapping

$$(x, y) \mapsto (-x, \sqrt{-1}y)$$

defines an order 4 automorphism α of E. This is automatically a group homomorphism and we have

$$\sigma(\alpha(T)) = \alpha(\sigma(T))$$

for all $T \in E_{tors}$ and all $\sigma \in \mathrm{Gal}(\overline{K}/K)$. For given n there is a matrix $A \in \mathrm{GL}_2(\mathbf{Z}/n\mathbf{Z})$, representing α, which commutes with all elements in the image of

$$\mathrm{Gal}(K(E[n])/K) \to \mathrm{GL}_2(\mathbf{Z}/n\mathbf{Z}).$$

If $n = p$ is a prime with $p \equiv 3 \mod 4$ we claim that A is not a scalar matrix. Let us suppose the contrary. As $A^4 = 1$ and since $\mathbf{F}_p^\times$ does not contain an element of order 4 we have $A^2 = 1$. Therefore, $\alpha^2(T) = T$ for some $T \in E[p]$ of order p. But $\alpha^2 = -1$ as an endomorphism on E, so $T = -T$ and this contradicts $p \neq 2$. So A is not a scalar matrix. The set of 2×2 matrices with coefficients in $\mathbf{F}_p$ that commute with A is a proper vector subspace. Thus it contains at most p^3 elements. But $\mathrm{GL}_2(\mathbf{F}_p)$ contains $(p^2 - 1)(p^2 - p)$ elements and there are infinitely many primes $p \equiv 3 \mod 4$. So we see that the index $[\mathrm{GL}_2(\mathbf{F}_p) : \operatorname{im} \rho_p]$ is unbounded as p varies.

An important theorem of Serre [33] states roughly that the image of ρ_n is almost everything as n varies if there are no exceptional endomorphisms as in the previous example. We say that E has complex multiplication or CM if, after possibly enlarging K, it admits a morphism $\alpha : E \to E$ with $\alpha(0) = 0$ that is not of the form $[n]$ for any $n \in \mathbf{Z}$.

Theorem 1.6 (Serre [33]) *Let us suppose that E does not have CM. There exists a constant $C > 0$, depending on E and K, such that*

$$[\mathrm{GL}_2(\mathbf{Z}/n\mathbf{Z}) : \operatorname{im} \rho_n] \leq C$$

for all $n \geq 1$.

Using Serre's result it is easy to find a lower bound for $[K(T):K]$ that is polynomial in terms of the order of a torsion point T.

Corollary 1.7 *Let E satisfy the hypothesis of the theorem above and let C be as in its conclusion. If $T \in E_{tors}$ has order n, then*

$$[K(T) : K] \geq \frac{6}{\pi^2 C} n^2.$$

Proof As above we can identify $E[n]$ with the group $(\mathbf{Z}/n\mathbf{Z})^2$ by finding T' such that (T, T') is a $\mathbf{Z}/n\mathbf{Z}$-basis of $E[n]$. Let us abbreviate $G = \operatorname{im} \rho_n \subseteq \mathrm{GL}_2(\mathbf{Z}/n\mathbf{Z})$. The index of G in $\mathrm{GL}_2(\mathbf{Z}/n\mathbf{Z})$ is at most C by Serre's theorem.
Say $\sigma \in \mathrm{Gal}(K(E[n])/K)$. Then $\sigma(T) = T$ if and only if

$$\rho_n(\sigma) = \begin{pmatrix} 1 & * \\ 0 & * \end{pmatrix} \in S = \left\{ \begin{pmatrix} 1 & b \\ 0 & d \end{pmatrix}; \ b \in \mathbf{Z}/n\mathbf{Z} \text{ and } d \in (\mathbf{Z}/n\mathbf{Z})^\times \right\}.$$

The group S is the stabilizer of $\binom{1}{0} \in (\mathbf{Z}/n\mathbf{Z})^2$ under the natural action of $\mathrm{GL}_2(\mathbf{Z}/n\mathbf{Z})$ on $(\mathbf{Z}/n\mathbf{Z})^2$. By Galois Theory we have $\mathrm{Gal}(K(E[n])/K(T)) \cong G \cap S$ and

$$[K(T):K] = [G:G \cap S] = \frac{\#G}{\#G \cap S} \geq \frac{\#G}{\#S} \geq \frac{1}{C}\frac{\#\mathrm{GL}_2(\mathbf{Z}/n\mathbf{Z})}{\#S}.$$

Certainly, $\#S = n\#(\mathbf{Z}/n\mathbf{Z})^{\times} = n\varphi(n) = n^2 \prod_{p|n}(1 - p^{-1})$, where p denotes a prime. It is known that $\#\mathrm{GL}_2(\mathbf{Z}/n\mathbf{Z}) = n^4 \prod_{p|n}(1 - p^{-2})(1 - p^{-1})$ and therefore

$$[K(T):K] \geq \frac{1}{C}n^2 \prod_{p|n}(1 - p^{-2}) \geq \frac{1}{C}n^2 \prod_{p}(1 - p^{-2}) = \frac{6}{C\pi^2}n^2,$$

as $\prod_p (1 - p^{-2})^{-1} = \sum_{m \geq 1} m^{-2} = \pi^2/6$. $\qquad\square$

This is the type of degree lower bound we are looking for. Up to a multiplicative constant it is best possible in view of the table above. However, it does not cover the case of elliptic curves with complex multiplication for which Serre's theorem does not apply. Luckily, in the CM case we have class field theory at our disposal. This powerful tool yields polynomial lower bounds in the CM case, even for abelian varieties, see for example work of Silverberg [35].

Using completely different methods inspired by Transcendence Theory, Masser obtained a polynomial lower bound for the degree. His main theorem [21] requires the Néron-Tate height which we will discuss in more detail below. It does not distinguish between elliptic curves with CM and those without CM.

Theorem 1.8 (Masser [21]) *For any elliptic curve E defined over a number field K there is a constant $c > 0$, depending on E and K, such that if $T \in E_{tors}$ has order n, then*

$$[K(T):K] \geq c\frac{n}{\log 2n}. \qquad (1.11)$$

Masser's lower bound (1.11) is linear up to a logarithmic term in n. For CM elliptic curves this cannot be improved to a quadratic lower bound as in Corollary 1.7.

David's Théorème 1.2 [8] makes the dependence of c in K explicit. It turns out that one can choose c to depend only on $[K : \mathbf{Q}]$ and that this dependency is polynomial. This feature is useful in some applications towards unlikely intersections that involve the Pila-Wilkie Theorem and its variants.

Further down we will describe a third approach to obtaining a polynomial lower bound for $[K(T) : K]$. It is based on intricate properties of the Néron-Tate on E.

3.2 The Height of an Algebraic Number

Before defining the height on an elliptic curve, we will state some facts on the absolute logarithmic Weil height of an algebraic number. This height can be defined in terms of the minimal polynomial over the integers.

Definition 1.9   Let $\alpha \in \overline{\mathbf{Q}}$ and let $P \in \mathbf{Q}[X]$ be an irreducible polynomial with coprime integral coefficients, $P(\alpha) = 0$, and leading coefficient $a > 0$. Then

$$h(\alpha) = \frac{1}{\deg P} \log \left(a \prod_{\substack{z \in \mathbf{C} \\ P(z)=0}} \max\{1, |z|\} \right)$$

is the absolute logarithmic Weil height, or just height, of α.

The conditions imposed on the polynomial P ensure that it is uniquely determined by α. Its degree $\deg P$ is just $[\mathbf{Q}(\alpha) : \mathbf{Q}]$.

Example 1.10

(i) *If $p, q \in \mathbf{Z}$ are coprime with $q > 0$, then $P = qX - p$ is the polynomial attached to p/q from the definition above. So*

$$h(p/q) = \log \max\{|p|, q\}.$$

(ii) *If $\alpha = 3 + \sqrt{2}$ then $P = X^2 - 6X + 7$ and so*

$$h(3 + \sqrt{2}) = \frac{1}{2} \log 7.$$

(iii) *For $\alpha = 1 + \sqrt{2}$ we find $P = X^2 - 2X - 1$ and*

$$h(1 + \sqrt{2}) = \frac{1}{2} \log(1 + \sqrt{2}).$$

(iv) *A root of unity ζ satisfies $\zeta^n = 1$ for some $n \in \mathbf{N}$. The polynomial P divides $X^n - 1$ in $\mathbf{Z}[X]$ and so its leading term is 1. Moreover, all conjugates of ζ lie on the unit circle. Therefore, $h(\zeta) = 0$.*

Let α be algebraic with $h(\alpha) \le B$. Up to a finite amount of ambiguity we may reconstruct P as in Definition 1.9 from B and $d = \deg P$ alone. Indeed, if we write

$$P = a_d X^d + \cdots + a_0 = a_d(X - \alpha_1) \cdots (X - \alpha_d),$$

with $\alpha_i \in \mathbf{C}$ and $a_0, \ldots, a_d \in \mathbf{Z}$ then we have the estimates

$$|a_0| = \left| a_d \prod_{i=1}^{d} \alpha_i \right| \le e^{dh(\alpha)} \le e^{dB}$$

$$|a_1| = \left| a_d \sum_{i=1}^{d} \prod_{j \neq i} \alpha_j \right| \leq d e^{dh(\alpha)} \leq d e^{dB}$$

$$\vdots$$

$$|a_i| \leq \binom{d}{i} e^{dB}.$$

where we used the triangle inequality to bound from above the $\binom{d}{i}$ terms appearing in a_i. This leaves us with only finitely many integers a_i for each i and the number of possibilities for P is at most

$$\prod_{i=0}^{d} \left(2 \binom{d}{i} e^{dB} + 1 \right) \leq \prod_{i=0}^{d} 3 e^{dB} \binom{d}{i} \leq 2^{d(d+1)} 3^{d+1} e^{Bd(d+1)}.$$

As P has at most d roots we obtain

$$\# \left\{ \alpha \in \overline{\mathbf{Q}}; \ h(\alpha) \leq B \text{ and } [\mathbf{Q}(\alpha) : \mathbf{Q}] \leq D \right\} \leq D^2 2^{D(D+1)} 3^{D+1} e^{BD(D+1)}.$$

$$(1.12)$$

We have just proved a quantitative version of

Theorem 1.11 (Northcott's Theorem) *Given real numbers B and D, the set of algebraic numbers of height at most B and degree at most D is finite.*

The estimate (1.12) leads to a very crude lower bound for $[\mathbf{Q}(\zeta) : \mathbf{Q}]$ for a root of unity ζ of order n. Indeed, the set in (1.12) contains n distinct elements $1, \zeta, \zeta^2, \ldots, \zeta^{n-1}$ if $B = 0$ and $d = [\mathbf{Q}(\zeta) : \mathbf{Q}]$. So $n \leq d^2 2^{d(d+1)} 3^{d+1}$, which is embarrassingly bad when compared to (1.10) as $\varphi(n) = d$. In any case, our estimate for d is not polynomial in n; it is useless in combination with the Pila-Wilkie Theorem. Further down, we will obtain a reasonable bound using more careful arguments involving the height.

There is an alternative definition of the height that makes certain computations more accessible.

Let K be a number field with ring of algebraic integers $\mathcal{O}_K$. To any element v of

$$M_K = \{\sigma : K \hookrightarrow \mathbf{C}\} / \{\text{complex conjugation}\}$$

$$\cup \{\mathcal{P}; \ \mathcal{P} \text{ is a non-zero prime ideal of } \mathcal{O}_K\}$$

we can attach an absolute value $| \cdot |_v : K \to [0, +\infty)$ in the following way. If v corresponds to a field embedding σ up to conjugation, we define $|\alpha|_v = |\sigma(\alpha)|$

for all $\alpha \in K$. We define the local degree

$$d_v = \begin{cases} 1 & : \text{if } \sigma(K) \subseteq \mathbf{R}, \\ 2 & : \text{else wise.} \end{cases}$$

For $v = \mathcal{P}$ a prime ideal of $\mathcal{O}_K$ we proceed as follows. The prime ideal $\mathcal{P} \cap \mathbf{Z}$ of $\mathbf{Z}$ is generated by a prime number p and $p\mathcal{O}_K = \mathcal{P}^e I$ for some ideal $I \subseteq \mathcal{O}_K$ that is not divisible by $\mathcal{P}$ and $e \in \mathbf{N}$. The exponent e is called the ramification index of $\mathcal{P}$. If $\alpha = 0$ we set of course $|\alpha|_v = 0$. If $\alpha \neq 0$ the fractional ideal $\alpha\mathcal{O}_K$ can be written as $\mathcal{P}^m J$ where J is a fractional ideal that does not involve $\mathcal{P}$ as a factor and $m \in \mathbf{Z}$. We set

$$|\alpha|_v = p^{-m/e}.$$

By definition we have $|p|_v = 1/p$. This normalization is convenient as $|\cdot|_v$ extends the usual p-adic absolute value on $\mathbf{Q}$. We attach the local degree

$$d_v = ef \quad \text{where} \quad [\mathcal{O}_K : \mathcal{P}] = p^f$$

to $v = \mathcal{P}$. This is none other than the degree of the completion of K with respect to $|\cdot|_v$ as an extension of the field of p-adic numbers. The absolute value $|\cdot|_v$ satisfies the ultrametric triangle inequality.

The principal fractional ideal generated by $\alpha \in K^\times$ factors into a finite product of prime ideals. So there are only finitely many $v \in M_K$ with $|\alpha|_v \neq 1$. Thus the sums in the proposition below are well defined.

Proposition 1.12 *If $\alpha \in K$, then*

$$h(\alpha) = \frac{1}{[K : \mathbf{Q}]} \sum_{v \in M_K} d_v \log \max\{1, |\alpha|_v\}. \tag{1.13}$$

If moreover $\alpha \neq 0$, then the product formula, stated here in logarithmic form,

$$\sum_{v \in M_K} d_v \log |\alpha|_v = 0 \tag{1.14}$$

holds true.

Proof For a proof of (1.13) we refer to Proposition 1.6.6 [5]. The product formula is a consequence of Proposition 1.4.4 [5]. $\qquad\qquad\square$

The right-hand side of (1.13) does not depend on the number field K containing α as the original definition of $h(\alpha)$ did not involve K.

This expression for the height opens up new avenues. For example, if $v \in M_K$ and $k \in \mathbf{N}$, then

$$\max\{1, |\alpha\beta|_v\} \le \max\{1, |\alpha|_v\} \max\{1, |\beta|_v\}$$
$$\text{and} \quad \max\{1, |\alpha^k|_v\} = \max\{1, |\alpha|_v\}^k$$

for all $\alpha, \beta \in K$. Taking the logarithm left and right, multiplying by d_v, summing over all $v \in M_K$, and dividing by $[K : \mathbf{Q}]$ yields

$$h(\alpha\beta) \le h(\alpha) + h(\beta) \quad \text{and} \quad h(\alpha^k) = kh(\alpha). \tag{1.15}$$

These statements hold true for all $\alpha, \beta \in \overline{\mathbf{Q}}$.

Using the equality in (1.15) together with Northcott's Theorem allows us to characterize algebraic numbers of the smallest possible height 0.

Theorem 1.13 (Kronecker's Theorem) *Let $\alpha \in \overline{\mathbf{Q}}^{\times}$. Then $h(\alpha) = 0$ if and only if α is a root of unity.*

Proof We already proved the "if" direction in Example 1.10(iv). Let us show the converse. We know $h(\alpha) = h(\alpha^2) = h(\alpha^3) = \cdots = 0$ by the equality in (1.15). By (1.12) there are at most finitely many pairwise distinct elements among the powers $\alpha, \alpha^2, \alpha^3, \ldots$. So $\alpha^i = \alpha^j$ for integers $i > j$. As $\alpha \ne 0$ we find $\alpha^{i-j} = 1$. So α is a root of unity. $\qquad\square$

Example 1.14 *Let ζ be a root of unity of order n taken as a complex number. As promised above we will use heights to find a polynomial lower bound for $[\mathbf{Q}(\zeta) : \mathbf{Q}]$ in terms of n. This bound will be asymptotically worse than the classical lower bound for $[\mathbf{Q}(\zeta) : \mathbf{Q}] = \varphi(n)$ given by (1.10). But it is polynomial in n and it is an illustrative warm up for the more difficult case of elliptic curves where no exact expression for the degree of a torsion point is available.*

We will work with an auxiliary integer m satisfying $2 \le m \le n$ and to be chosen properly as a function of n later on. Let us write $z_k = e^{2\pi\sqrt{-1}k/n}$ for $k \in \mathbf{Z}$; our root of unity is one of these values and all z_k are in $\mathbf{Q}(\zeta) \subseteq \mathbf{C}$. Now

$$|z_l - z_k| = \left|1 - e^{2\pi\sqrt{-1}(k-l)/n}\right| \le 2\pi\frac{|k-l|}{n} \le 2\pi\frac{m-1}{n}$$

if $0 \leq k, l < m$. *We use these inequalities to bound*

$$\delta = \prod_{\substack{0 \leq k,l < m \\ k \neq l}} (z_l - z_k)$$

from above by

$$|\delta| \leq \left(2\pi \frac{m-1}{n} \right)^{m(m-1)}. \tag{1.16}$$

But δ is an element of the number field $K = \mathbf{Q}(\zeta)$. If $v \in M_K$ comes from an embedding $\sigma : K \hookrightarrow \mathbf{C}$, then we can express δ as a Vandermonde determinant, i.e.

$$|\delta|_v = \left| \prod_{\substack{0 \leq k,l < m \\ k \neq l}} \sigma(z_l) - \sigma(z_k) \right|$$

$$= \left| \det \begin{pmatrix} 1 & \sigma(z_0) & \sigma(z_0)^2 & \cdots & \sigma(z_0)^{m-1} \\ \vdots & \vdots & \vdots & & \vdots \\ 1 & \sigma(z_{m-1}) & \sigma(z_{m-1})^2 & \cdots & \sigma(z_{m-1})^{m-1} \end{pmatrix} \right|.$$

By Hadamard's inequality we obtain $|\delta|_v \leq m^{m/2}$ since $|\sigma(z_i)| = 1$ for all i. If v is induced by the inclusion $K \subseteq \mathbf{C}$ we get a better bound (1.16), at least if m is small compared to n.

If $v \in M_K$ is a prime ideal of $\mathcal{O}_K$, then $|\delta|_v \leq 1$ due to the fact that each z_i is an algebraic integer.

No factor in δ vanishes, so $\delta \neq 0$. By the product formula we have

$$0 = \sum_{v \in M_K} d_v \log |\delta|_v \leq m(m-1) \log \left(2\pi \frac{m-1}{n} \right) + [K : \mathbf{Q}] \frac{m}{2} \log m.$$

We divide this expression by $(m \log m)/2$ and rearrange terms to obtain

$$2 \frac{m-1}{\log m} \log \left(\frac{n}{2\pi(m-1)} \right) \leq [K : \mathbf{Q}].$$

With the choice $m = 1 + \max\{1, [n/7]\} \geq 2$ we get the estimate

$$c \frac{n}{\log 2n} \leq [K : \mathbf{Q}] = [\mathbf{Q}(\zeta) : \mathbf{Q}]$$

for some absolute constant $c > 0$.

We observe that the height has almost completely disappeared from the argument above. Implicitly, we used that each local contribution $\log \max$

$\{1, |\zeta|_v\}$ *vanishes when we bounded* $|\delta|_v$ *from above. The analogous statement turns out to be false for torsion points on elliptic curves.*

This strategy can be used to bound from above the number of points of small height, instead of just height 0, *in a number field. See for example the work of Loher and Masser [19] which also uses Hadamard's inequality to bound a Vandermonde determinant.*

3.3 The Néron-Tate Height on Elliptic Curves

We see from (1.15) that the absolute logarithmic Weil height plays nicely with the multiplicative structure of $\overline{\mathbf{Q}}^{\times}$.

Let E be an elliptic curve given by a Weierstrass equation (1.2) with coefficients in a number field $K \subseteq \overline{\mathbf{Q}}$. The naive height of an algebraic point of E is defined as follows.

Definition 1.15 Say $P \in E(\overline{\mathbf{Q}})$.

(i) If $P = (x, y)$ lies on the affine part of E, then

$$h_{\text{naive}}(P) = \frac{1}{2}h(x).$$

(ii) If $P = 0 = [0 : 1 : 0]$ is the point at infinity, then $h_{\text{naive}}(P) = 0$.

The factor $1/2$ in the definition of the naive height is a convention motivated by the fact that the x-coordinate is a rational function of degree 2 on E. The absence of the y-coordinate from the definition of the naive height is harmless as its value is determined up to sign by the x-coordinate.

Example 1.16

(i) *Say E is given by $y^2 = 4x^3 - 4x$, as in Example 1.5 where we computed the group $E[2]$. We have*

$$h_{\text{naive}}(0) = 0, \quad h_{\text{naive}}((0,0)) = \frac{1}{2}h(0) = 0, \quad \text{and} \quad h_{\text{naive}}((\pm 1, 0)) = 0.$$

(ii) *Now suppose that E is given by $y^2 = 4x^3 - 1$. Any root of $4X^3 - 1$ yields the x-coordinate of a point on $E(\overline{\mathbf{Q}})$ of order 2. For example, $(4^{-1/3}, 0)$ has order 2. The naive height of this point is*

$$h_{\text{naive}}((4^{-1/3}, 0)) = \frac{1}{2}h(4^{-1/3}) = \frac{1}{6}\log 4.$$

The naive height of a 2-torsion point vanishes in (i). But this is a red herring. The second example suggests that there is no numerical characterization of torsion points on E in terms of the naive height similar to Kronecker's Theorem above. The canonical or Néron-Tate height will rectify this problem.

Tate proposed an elegant construction of a height using an iterative construction involving the naive height. To jump start this construction we require a technical lemma which describes the growth of the naive height under the duplication morphism.

Lemma 1.17 *There exists a constant $C > 0$ depending only on E such that*

$$|h_{\text{naive}}([2](P)) - 4h_{\text{naive}}(P)| \leq C$$

for all $P \in E(\overline{\mathbf{Q}})$.

Proof The claim is certainly true for a sufficiently large C for the four points in $E[2]$. Let us suppose $[2](P) \neq 0$ in which case we may write $P = (x, y)$ with x in some number field $F \supseteq K$. We have $g(x) \neq 0$ where $g = 4X^3 - g_2 X - g_3$ and so by (1.3) also

$$[2](P) = \left(\frac{f(x)}{g(x)}, *\right)$$

where $f = X^4 + g_2 X^2/2 + 2g_3 X + g_2^2/16$.

We use (1.13) and (1.14) applied to $g(x) \neq 0$ to rewrite

$$2h_{\text{naive}}([2](P)) = h(f(x)/g(x))$$
$$= \frac{1}{[F:\mathbf{Q}]} \sum_{v \in M_F} d_v \log \max\{1, |f(x)/g(x)|_v\}$$
$$= \frac{1}{[F:\mathbf{Q}]} \sum_{v \in M_F} d_v \log \max\{|g(x)|_v, |f(x)|_v\}.$$

Let $v \in M_F$. We note that g has degree 3 and f has degree 4 in X. Using the triangle inequality if v is a field embedding and the ultrametric triangle inequality if v is a prime ideal we estimate

$$\max\{|g(x)|_v, |f(x)|_v\} \leq c_v \max\{1, |x|_v\}^4;$$

here and below $c_v > 0$ denote constants which are independent of x and equal 1 for all but finitely many $v \in M_F$. Taking the logarithm, multiplying by d_v, summing over all $v \in M_F$, and dividing by $2[F:\mathbf{Q}]$ yields

$$h_{\text{naive}}([2](P)) \leq 4h_{\text{naive}}(P) + C$$

where C depends only on E. This is half of the assertion.

The other half, i.e. the corresponding lower bound for $h_{\text{naive}}([2](P))$ is more difficult as we cannot apply the triangle inequality directly. We will need the fact that E is smooth, i.e. $\Delta = g_2^3 - 27g_3^2 \neq 0$. One verifies the polynomial identity

$$2^8 \Delta X^7 = af + bg$$

where

$$
\begin{aligned}
a &= 256(g_2^3 - 27g_3^2)X^3 - 64g_2^2g_3X^2 + (1408g_2g_3^2 - 48g_2^4)X \\
&\quad - 48g_2^3g_3 + 1536g_3^3, \\
b &= 16g_2^2g_3X^3 + (-20g_2^4 + 512g_2g_3^2)X^2 + (-104g_2^3g_3 + 3072g_3^3)X \\
&\quad + 96g_2^2g_3^2 - 3g_2^5
\end{aligned}
$$

both have degree 3 in X. As above, we apply the triangle or ultrametric triangle inequality to estimate

$$|\Delta|_v |x|_v^7 \leq c_v \max\{1, |x|_v\}^3 \max\{|f(x)|_v, |g(x)|_v\} \tag{1.17}$$

where $c_v > 0$ is a new sequence of constants which satisfies the same finiteness property as before. Similar estimates applied to the identity

$$\Delta = -16(3X^2 - g_2)f + (12X^3 + 5g_2X + 27g_3)g$$

give

$$|\Delta|_v \leq c_v \max\{1, |x|_v\}^3 \max\{|f(x)|_v, |g(x)|_v\}$$

after modifying the c_v if necessary. This bound and (1.17) can be stated simultaneously as

$$|\Delta|_v \max\{1, |x|_v\}^7 \leq c_v \max\{1, |x|_v\}^3 \max\{|f(x)|_v, |g(x)|_v\}.$$

We cancel $\max\{1, |x|_v\}^3$, take the logarithm, multiply by the local factors d_v, and sum over all $v \in M_F$. The contribution coming from $|\Delta|_v$ sums up to zero by the product formula (1.14). Finally, we divide by $[F : \mathbf{Q}]$ to obtain

$$8h_{\text{naive}}(P) = 4h(x) \leq 2C + \frac{1}{[F : \mathbf{Q}]} \sum_{v \in M_F} d_v \log \max\{|f(x)|_v, |g(x)|_v\} \tag{1.18}$$

after possibly increasing C. We apply the product formula a last time to see that the right-hand side of (1.18) equals $2C + h(f(x)/g(x)) = 2C + 2h_{\text{naive}}([2](P))$. So $4h_{\text{naive}}(P) \leq h_{\text{naive}}([2](P)) + C$ which was the remaining claim. $\square$

Now we can complete Tate's procedure for the construction of the Néron-Tate height on $E(\overline{\mathbf{Q}})$.

Proposition-Definition 1.18 *For any $P \in E(\overline{\mathbf{Q}})$ the limit*

$$\lim_{n \to +\infty} \frac{h_{\text{naive}}([2^n](P))}{4^n}$$

exists and is called the Néron-Tate or canonical height of P. It is denoted by $\hat{h}(P)$ and satisfies $\hat{h}([2](P)) = 4\hat{h}(P)$. Moreover, $|\hat{h}(P) - h_{\text{naive}}(P)| \le C/3$ with C the constant from Lemma 1.17.

Proof We elaborate here this classical argument which involves the telescoping sum

$$\frac{h_{\text{naive}}([2^n](P))}{4^n} - \frac{h_{\text{naive}}([2^m](P))}{4^m}$$

$$= \sum_{k=m}^{n-1} \frac{h_{\text{naive}}([2^{k+1}](P)) - 4h_{\text{naive}}([2^k](P))}{4^{k+1}},$$

for integers $n \ge m \ge 0$. So

$$\left| \frac{h_{\text{naive}}([2^n](P))}{4^n} - \frac{h_{\text{naive}}([2^m](P))}{4^m} \right| \qquad (1.19)$$

$$\le \sum_{k=m}^{n-1} \frac{|h_{\text{naive}}([2^{k+1}](P)) - 4h_{\text{naive}}([2^k](P))|}{4^{k+1}}$$

$$\le \sum_{k=m}^{n-1} \frac{C}{4^{k+1}}$$

$$\le \frac{C}{4^{m+1}} \sum_{k=0}^{\infty} \frac{1}{4^k}$$

$$= \frac{C}{3 \cdot 4^m}$$

where we applied Lemma 1.17 to $[2^k](P)$ to get the second inequality. Therefore,

$$\left(\frac{h_{\text{naive}}([2^n](P))}{4^n} \right)_{n \ge 1}$$

is a Cauchy sequence and from this we deduce our claim on convergence.

The remaining assertions are now almost formal. Indeed, we shift by 2 and get

$$\hat{h}([2](P)) = \lim_{n \to +\infty} \frac{h_{\text{naive}}([2^{n+1}](P))}{4^n} = 4 \lim_{n \to +\infty} \frac{h_{\text{naive}}([2^{n+1}](P))}{4^{n+1}} = 4\hat{h}(P).$$

On setting $m = 0$ and taking $n \to +\infty$ in (1.19) we find $|\hat{h}(P) - h_{\text{naive}}(P)|$ $\leq C/3$. $\qquad\square$

Example 1.19 *The order of $T \in E_{tors}$ is at least the order of any sequence member*

$$T, \quad [2](T), \quad [2^2](T), \quad [2^3](T), \quad \ldots.$$

So this sequence attains only finitely many values. In particular, $[2^m](T) = [2^n](T)$ for integers $0 \leq m < n$. The Néron-Tate height, which satisfies $\hat{h}([2^m]) = 4^m \hat{h}(T)$ and $\hat{h}([2^n]) = 4^n \hat{h}(T)$ by the proposition-definition above, yields $\hat{h}(T) = 0$.

One can go further and prove $\hat{h}([n](P)) = n^2 \hat{h}(P)$ for all $P \in E(\overline{\mathbf{Q}})$, see Chapter VIII, Theorem 9.3 [36]. The Néron-Tate height is a quadratic function and this makes proving that torsion points have height 0 even simpler.

Corollary 1.20 *There are only finitely many K-rational torsion points, i.e. $E_{tors} \cap E(K)$ is a finite group.*

Proof The Néron-Tate height of any $T \in E_{\text{tors}} \cap E(K)$ vanishes. So $h_{\text{naive}}(T) \leq C/3$ by Proposition-Definition 1.18. If $T \neq 0$, then $T = (x, y)$ with $h(x) \leq 2C/3$. But x is in the number field K. By Northcott's Theorem there are only finitely many possibilities for x and at most twice as many for the pair (x, y). $\qquad\square$

We could now deduce a lower bound for $[K(T) : K]$ in terms of the order of T by combining the argument in the corollary above with the argument given just below Northcott's Theorem 1.11. However, the bound one gets is not polynomial and thus not good enough for our application.

Kronecker's Theorem also holds for the Néron-Tate height.

Theorem 1.21 (Kronecker's Theorem for the Néron-Tate height) *Say $P \in E(\overline{\mathbf{Q}})$. Then $\hat{h}(P) = 0$ if and only if $P \in E_{tors}$.*

Proof The "if" direction was proved in the example above. The "only if" implication follows as in the multiplicative setting, cf. Theorem 1.13. Observe that Northcott's Theorem holds true for the Néron-Tate height by the argument given in the proof of Corollary 1.20. $\qquad\square$

3.4 The Local Decomposition of the Néron-Tate Height

In order to mimic the argument in Example 1.14 we require a decomposition
of the Néron-Tate height into local terms much as (1.13) is a decomposition
of the Weil height into local terms. We state below, without proof, the Local
Decomposition Theorem attributed to Néron and Tate.

Let E be an elliptic curve given by (1.2) with g_2, g_3 in some number field K.
For any $v \in M_K$ we define a complete field K_v as follows. If v corresponds to
an embedding $K \hookrightarrow \mathbf{R}$ we set $K_v = \mathbf{R}$, if v comes from a non-real embedding
$K \hookrightarrow \mathbf{C}$, then $K_v = \mathbf{C}$. If v is a prime ideal of $\mathcal{O}_K$ we set K_v to be the
completion of K with respect to the absolute value $| \cdot |_v$. The group of K_v-
rational points of E is denoted by $E(K_v)$.

Theorem 1.22 *Let $\Delta = g_2^3 - 27g_3^2 \neq 0$ be the discriminant of (1.2). For any
$v \in M_K$ there exists a continuous function*

$$\lambda_v : E(K_v) \smallsetminus \{0\} \to \mathbf{R}$$

with the following properties for all $P = (x, y)$ in the affine part of $E(K_v)$.

(i) *We have*

$$\lambda_v(x, y) = \frac{1}{2} \log |x|_v + O_v(1)$$

*on any v-adic neighborhood of $\{0\}$, the constant implicit in $O_v(1)$ may
depend on the neighborhood but not on the point (x, y).*

(ii) *If $[2](P) \neq 0$, then*

$$\lambda_v([2](P)) = 4\lambda_v(P) - \log |y|_v + \frac{1}{4} \log |\Delta|_v. \qquad (1.20)$$

(iii) *If v corresponds to a prime ideal of $\mathcal{O}_K$ with $g_2, g_3 \in \mathcal{O}_K$, $|2|_v = 1$, and
$|\Delta|_v = 1$, then*

$$\lambda_v(x, y) = \frac{1}{2} \log \max\{1, |x|_v\} \geq 0$$

for all $(x, y) \in E(K_v) \smallsetminus \{0\}$.

Finally, if $P \in E(K) \smallsetminus \{0\}$, then

$$\hat{h}(P) = \frac{1}{[K : \mathbf{Q}]} \sum_{v \in M_F} d_v \lambda_v(P). \qquad (1.21)$$

Proof We refer to Theorem 1.1, Chapter VI [37] for a proof of the existence
of the λ_v and properties (i) and (ii). We must divide (1.2) by 4 to obtain a
Weierstrass equation as in this reference. This is the reason for the discrepancy

between the term $\log |y|_v$ in (1.20) and the corresponding term in (iii) of *loc. cit.* Our part (iii) follows from Remark 4.1.1 *loc. cit.* as the condition implies that we have good reduction at v. The local decomposition (1.21) is proved in Theorem 2.1 *loc. cit.* □

Remark 1.23

(i) For a given $P \in E(K) \setminus \{0\}$ the sum in (1.21) involves only finitely many non-zero terms. Indeed, for all but finitely many v the condition given in part (iii) of the theorem is met and hence $\lambda_v(P) = 0$.

(ii) Each function λ_v is uniquely determined by the properties stated in (i) and (ii) of the theorem and from the fact that $\lambda_v : E(K_v) \setminus \{0\} \to \mathbf{R}$ is continuous. This statement is part of Theorem 1.1 cited above from Silverman's book.

(iii) It is possible that λ_v takes negative values. This can only happen for finitely many v: those corresponding to a field embedding or those that do not satisfy the condition in part (iii) of the theorem above. In any case, λ_v is bounded from below by a constant on $E(K_v) \setminus \{0\}$; indeed, just apply (i) from the theorem and use the continuity of λ_v.

To bound $[K(T) : K]$ from below for a torsion point T we follow the strategy given in Example 1.14. We replace Hadamard's inequality by the following theorem which is due to Elkies in the archimedean case.

Theorem 1.24 *Let $P_1, \ldots, P_n \in E(K)$ be pairwise distinct points and suppose $v \in M_K$. There exists a constant $C \geq 0$ depending only on E and v with the following property.*

(i) *If v comes from an embedding of K into $\mathbf{C}$, then*

$$\sum_{\substack{1 \leq i,j \leq n \\ i \neq j}} \lambda_v(P_i - P_j) \geq -\frac{n}{2} \log n - Cn. \tag{1.22}$$

(ii) *If v is a prime ideal, then*

$$\sum_{\substack{1 \leq i,j \leq n \\ i \neq j}} \lambda_v(P_i - P_j) \geq -Cn. \tag{1.23}$$

Proof The deeper archimedean estimate is due originally to Elkies, see Theorem 5.1 in Chapter VI of Lang's book [18] or Hriljac's paper [15]. Baker and Petsche gave a proof of (i) and (ii) with explicit constants, cf. Propositions A.1 and 3.4 [2]. □

We note that by Theorem 1.22(iii) the sum (1.23) is non-negative if $g_2, g_3 \in \mathcal{O}_F$, $|2|_v = 1$, and $|\Delta|_v = 1$.

In Appendix we give a proof of the archimedean estimate in (i) if the period lattice of E with respect to v is rectangular. Our argument is based on the functional equation for the classical theta function. It avoids direct reference to properties of the solution of the heat equation such as the maximum principle.

The $\frac{n}{2} \log n$ term in the archimedean case is reminiscent of $m^{m/2}$ coming out of Hadamard's inequality in Example 1.14. A crucial feature of both (1.22) and (1.23) is that the lower bounds are asymptotically better than n^2. We now use these lower bounds to estimate the size of the Galois orbit of a torsion point.

Theorem 1.25   *There exists a constant $c > 0$ which depends only on E such that if $T \in E_{tors}$ has order n, then*

$$[K(T) : K] \geq c \frac{n}{\log 2n}.$$

Proof   Say $T \in E(F)$ with F a number field containing K. We may assume $F \subseteq \mathbf{C}$. Let $v_0 \in M_F$ be determined by the inclusion $F \hookrightarrow \mathbf{C}$ and let $C \geq 1$ be an integer, to be determined later on.

Any point in $E(\mathbf{C}) \smallsetminus \{0\}$ corresponds via (1.9) to a point in $(-1/2, 1/2]^2$; of course, 0 corresponds to $(0, 0)$.

We split $(-1/2, 1/2]^2$ up into C^2 half-open squares of side length $1/C$. By the Pigeonhole Principle at least n/C^2 of the n points

$$0, T, [2](T), \ldots, [n-1](T)$$

have image in a single square. We write $z_1, \ldots, z_m \in (-1/2, 1/2]^2$ for these images where m is the least integer greater than or equal to n/C^2. So $|z_i - z_j| \leq 1/C$ for the sup-norm on $\mathbf{R}^2$. Let $P_1, \ldots, P_m$ denote the distinct arguments.

The Weierstrass function has a double pole at lattice points. This combined with Theorem 1.22(i) implies $\lambda_{v_0}(P_i - P_j) \geq \frac{1}{2} \log C$ if C is sufficiently large. By the Local Decomposition Theorem we have

$$\sum_{\substack{1 \leq i,j \leq m \\ i \neq j}} \hat{h}(P_i - P_j) = \frac{1}{[F:\mathbf{Q}]} \sum_{v \in M_F} \sum_{\substack{1 \leq i,j \leq m \\ i \neq j}} d_v \lambda_v(P_i - P_j)$$

$$\geq m(m-1) \frac{\log C}{2[F:\mathbf{Q}]}$$

$$+ \frac{1}{[F:\mathbf{Q}]} \sum_{\substack{v \in M_F \\ v \neq v_0}} \sum_{\substack{1 \leq i,j \leq m \\ i \neq j}} d_v \lambda_v(P_i - P_j).$$

We apply Theorem 1.24 to bound the remaining places from below

$$\sum_{\substack{1 \leq i,j \leq m \\ i \neq j}} \hat{h}(P_i - P_j) \geq m(m-1)\frac{\log C}{2[F:\mathbf{Q}]} - cm\log m \qquad (1.24)$$

where $c \geq 1$ depends only on E; observe that $\lambda_v(P_i - P_j)$ is non-negative for all but finitely many v by the argument just below Theorem 1.24.

Each P_i has finite order and so does $P_i - P_j$. Therefore, the Néron-Tate heights in the sum on the left in (1.24) vanish. Rearranging the terms yields

$$[F:\mathbf{Q}] \geq \frac{\log C}{2c}\frac{m-1}{\log m}.$$

The theorem follows as $m \geq n/C^2$ and since $t \mapsto (t-1)/\log t$ is increasing on $(0, +\infty)$. $\qquad \square$

Hindry and Silverman [13] used an explicit version of the inequality of Elkies to prove the uniform bound

$$\#E_{\text{tors}} \cap E(K) \leq 1977408[K:\mathbf{Q}]\log[K:\mathbf{Q}]$$

if E has good reduction at all finite places of K. Merel [23] proved earlier and with entirely different methods that $\#E_{\text{tors}} \cap E(K)$ is bound in function of $[K:\mathbf{Q}]$ only for unrestricted E after Mazur [22] did this for $K = \mathbf{Q}$. Merel's bound is not polynomial in $[K:\mathbf{Q}]$, however.

4 Application to the Manin-Mumford Conjecture

We now sketch a proof of Theorem 1.2 following the strategy used in Pila and Zannier's paper [27].

As mentioned in the introduction, the "if" direction is classical. We will only prove the "only if" direction. So let us assume that the curve $\mathcal{X}$ contains infinitely many torsion points of E^g.

We begin by making some reduction steps.

Certainly, $g \geq 1$ and there is nothing to show if we have equality. We may thus assume $g \geq 2$.

For $1 \leq i < j \leq g$ we have projections $\pi_{ij}: E^g \to E^2$ onto the i-th and j-th factors of E^g. If a restriction $\pi_{ij}|_{\mathcal{X}}$ happens to be constant, then its value must be a torsion point. If it is not constant, then it maps $\mathcal{X}$ to a curve in E^2 containing infinitely many torsion points. The preimage of an algebraic subgroup under π_{ij} is an algebraic subgroup of E^g. From this we see that it suffices to prove the theorem if $g = 2$.

The argument now proceeds in four steps.

Step 0 (Setting-up the definable set) We uniformize E as in Section 2.2. The cartesian square of the elliptic logarithm ξ defined near (1.9) yields a function

$$\Xi : (E(\mathbf{C}) \smallsetminus \{0\})^2 \to (-1/2, 1/2]^4$$

which is definable in $\mathbf{R}_{an}$. If $\mathcal{X}$ has infinite intersection with $E_{tors} \times \{0\}$ or $\{0\} \times E_{tors}$ then it has the form $E \times \{0\}$ or $\{0\} \times E$ in which case our claim holds. So let

$$\mathcal{Z} = \Xi \left(\mathcal{X}(\mathbf{C}) \cap (E(\mathbf{C}) \smallsetminus \{0\})^2 \right).$$

This set is contained in $\mathbf{R}^4$, definable in $\mathbf{R}_{an}$, and contains infinitely many rational points.

Step 1 (Lower bounds for the Galois orbit) If $T \in \mathcal{X}(\mathbf{C})$ is a torsion point of order $n \geq 2$, then

$$\Xi(T) \in \frac{1}{n}\mathbf{Z}^4 \cap (-1/2, 1/2]^4.$$

So $H(\Xi(T)) \leq n$, where

$$H(r_1, \ldots, r_4) = \max\{e^{h(r_1)}, \ldots, e^{h(r_4)}\}$$

is the exponential height for any rational vector $(r_1, \ldots, r_4) \in \mathbf{Q}^4$.

We saw in Section 2.1 that the torsion points of E are algebraic over K. Moreover, any conjugate $\sigma(T)$ with $\sigma \in \mathrm{Gal}(\overline{K}/K)$ is again torsion. It yields a new rational point $\Xi(\sigma(T))$, again with height at most n. According to Theorem 1.25 the size of the Galois orbit is at least $cn/\log 2n$ where $c > 0$ may depend on E but not on n. As the curve $\mathcal{X}$ is the zero set of polynomials with coefficients in K the torsion point $\sigma(T)$ lies on $\mathcal{X}$ and thus $\Xi(\sigma(T)) \in \mathcal{Z}$. Therefore,

$$\#\left\{ r \in \mathcal{Z} \cap \mathbf{Q}^4; \ H(r) \leq n \right\} \geq c\frac{n}{\log 2n}. \tag{1.25}$$

Step 2 (Counting rational points using the Pila-Wilkie Theorem) The lower bound (1.25) competes with the upper bound coming from the Theorem of Pila-Wilkie [26], discussed in Alex Wilkie's course. Say $\epsilon > 0$ and let $\mathcal{Z}^{alg}$ be the union of all connected, real semi-algebraic curves contained in $\mathcal{Z}$. Then this theorem implies

$$\#\left\{ r \in (\mathcal{Z} \smallsetminus \mathcal{Z}^{alg}) \cap \mathbf{Q}^4; \ H(r) \leq n \right\} \leq C(\epsilon)n^\epsilon \tag{1.26}$$

where $C(\epsilon)$ may depend on $\mathcal{Z}$ and ϵ but not on n.

Any fixed $\epsilon < 1$, such as $\epsilon = 1/2$, will do. Indeed, $\mathcal{X}(\mathbf{C}) \cap E^2_{\text{tors}}$ contains infinitely many torsion points by hypothesis and E^2_{tors} contains at most finitely many of fixed order by (1.5). Therefore, we may suppose that n satisfies

$$C(1/2)n^{1/2} < c\frac{n}{\log 2n}.$$

Assuming this inequality, (1.25) and (1.26) together imply $Z^{\text{alg}} \neq \emptyset$.

Step 3 (Applying the Ax-Lindemann-Weierstrass Theorem) In step 2 we showed that Z contains a real semi-algebraic curve. So there is a real semi-algebraic and real analytic, non-constant function $(-1/2, 1/2) \rightarrow Z$ that parametrizes this curve. The four coordinate functions have a Taylor expansion in a neighborhood of 0. The Taylor series converge on an open neighborhood of 0 in $\mathbf{C}$. We thus obtain 4 holomorphic maps, each with target $\mathbf{C}$, in a neighborhood of 0. On the reals, each map describes a coordinate with respect to a period lattice basis attached to one factor of E^2. As the period lattice lies inside $\mathbf{C}$ we may also allow complex coordinates. Taking the linear combination with respect to the period lattice basis we end up with two non-constant, holomorphic maps γ_1, γ_2 defined on a neighborhood of $0 \in \mathbf{C}$. They are algebraically dependent as the image of the original map is a real semi-algebraic curve. We write $\wp$ for the Weierstrass function. The transcendence degree satisfies

$$\text{trdeg } \mathbf{C}(\gamma_1, \gamma_2, \wp \circ \gamma_1, \wp \circ \gamma_2)/\mathbf{C} \leq \text{trdeg } \mathbf{C}(\gamma_1, \gamma_2)/\mathbf{C}$$
$$+ \text{trdeg } \mathbf{C}(\wp \circ \gamma_1, \wp \circ \gamma_2)/\mathbf{C} \quad (1.27)$$
$$\leq 1 + 1 = 2,$$

where $\text{trdeg } \mathbf{C}(\gamma_1, \gamma_2)/\mathbf{C} \leq 1$ comes from the semi-algebraic curve and $\text{trdeg } \mathbf{C}(\wp \circ \gamma_1, \wp \circ \gamma_2)/\mathbf{C} \leq 1$ holds since $\wp \circ \gamma_1$ and $\wp \circ \gamma_2$ are coordinates on the algebraic curve $\mathcal{X}$.

Enter the Ax-Lindemann-Weierstrass Theorem [1] for the Weierstrass function. See Orr's notes in this volume. It is a variant of the statement discussed in Jonathan Pila's lecture course for the classical exponential function. See also Brownawell and Kubota's [6] and Pila's proof [25] using o-minimality. The additional algebraic relation, witnessed numerically by (1.27), must be due to the existence of a linear relation

$$a\gamma_1 + b\gamma_2 + c = 0$$

between the functions $\gamma_{1,2}$ where $a, b, c \in \mathbf{C}$ are constants and $(a, b) \neq 0$. The existence of this relation forces the difference $\mathcal{X} - \mathcal{X}$, whose complex points are

$$\{P - Q; \ P, Q \in \mathcal{X}(\mathbf{C})\},$$

to be a proper subvariety of E^2. So $\mathcal{X} - \mathcal{X}$ is again an irreducible curve as is the sum $(\mathcal{X}-\mathcal{X})+(\mathcal{X}-\mathcal{X})$ by a similar argument. So $(\mathcal{X}-\mathcal{X})+(\mathcal{X}-\mathcal{X}) = \mathcal{X}-\mathcal{X}$ and we conclude that $\mathcal{X} - \mathcal{X}$ is a connected algebraic subgroup of E^2. We fix any $T \in \mathcal{X}(\mathbf{C})$ with finite order, then $\mathcal{X}$ is the translate of the said group by T. So $\mathcal{X}$ is an irreducible component of an algebraic subgroup. This is what we set out to prove. $\square$

Appendix   An Inequality of Elkies for the Local Néron-Tate Height

Let E be an elliptic curve presented by (1.2) with $g_2, g_3 \in K$ and K a number field. For an embedding $\sigma : K \hookrightarrow \mathbf{C}$ we have the local Néron-Tate height λ : $E(\mathbf{C}) \setminus \{0\} \to \mathbf{R}$ by Theorem 1.22; it is harmless to assume that K has no real embeddings so the completion is indeed $\mathbf{C}$. We will treat K as a subfield of $\mathbf{C}$. Our aim in this section is to prove Theorem 1.24(i) for a particular class of elliptic curves E. To an extent we follow the exposition in Baker and Petsche's work [2]. However, our argument differs slightly as it relies on transformation properties of the classical theta function and does not make reference to the heat equation. On the other hand, our approach is restricted to elliptic curves with j-invariant lying in $[1728, +\infty)$. We assume this hypothesis throughout the section, it forces the period lattice to be rectangular. We will prove the following proposition.

Proposition 1.26   *If $P_1, \ldots, P_n \in E(\mathbf{C})$ are pair-wise distinct points, then*

$$\sum_{\substack{1 \le i,j \le n \\ i \ne j}} \lambda(P_i - P_j) \ge -\frac{n}{2} \log n - Cn$$

where C is a constant which depends only on E.

After a change of variables we may assume that the Weierstrass function $\wp$ is associated to $\Omega = \mathbf{Z} + \tau\mathbf{Z}$ for some τ in $\mathbf{H}$, the upper half-plane in $\mathbf{C}$. We may assume that τ lies in the classical fundamental domain for the action of $\mathrm{SL}_2(\mathbf{Z})$ on $\mathbb{H}$. We also write λ for λ composed with $z \mapsto (\wp(z), \wp'(z))$. Thus λ is real analytic on $\mathbf{C} \setminus \Omega$ and Ω-periodic, see for example Theorem 3.2 in Chapter VI [37].

We start out as in Baker and Petsche's paper [2] and develop λ as a Fourier series in

$$\psi_\omega(z) = e^{2\pi\sqrt{-1}\langle \omega, z \rangle} \quad \text{where} \quad \langle \omega, z \rangle = au + bv$$

using the notation $\omega = a + b\tau \in \Omega$ with $a, b \in \mathbf{Z}$ and $z = u + v\tau \in \mathbf{C}$ with $u, v \in \mathbf{R}$. Indeed,

$$\lambda(z) = \frac{\text{Im}(\tau)}{2\pi} \sum_{\omega \in \Omega \smallsetminus \{0\}} \frac{1}{|\omega'|^2} \psi_\omega(z) \tag{1.28}$$

where

$$\omega' = b - a\tau \quad \text{if} \quad \omega = a + b\tau \tag{1.29}$$

determines an automorphism of Ω.

A priori we must interpret the equality in (1.28) as follows. In the coordinates u and v we may understand each ψ_ω as an element of the Hilbert space $L^2([0,1)^2)$. The right-hand side of (1.28) converges to λ in $L^2([0,1)^2)$. The question of *pointwise convergence* will keep us busy for some time.

As Ω contains roughly r^2 elements of modulus at most r we observe that $\sum_{\omega \in \Omega \smallsetminus \{0\}} |\omega'|^{-2} = +\infty$. So the series on the right in (1.28) does not converge absolutely for any argument, and it does not converge at all at $z = 0$.

Elkies's insight was to dampen the Fourier coefficients by convoluting with the heat kernel. We will not specify what this means but rather define explicitly

$$\lambda_t(z) = \frac{\text{Im}(\tau)}{2\pi} \sum_{\omega \in \Omega \smallsetminus \{0\}} \frac{1}{|\omega'|^2} e^{-\frac{2\pi}{\text{Im}(\tau)} |\omega'|^2 t} \psi_\omega(z) \tag{1.30}$$

for any $t > 0$ where we retain the notation (1.29). We use S to denote the value of the finite sum $\sum_{\omega \in \Omega, 0 < |\omega'| < 1} |\omega'|^{-2} e^{-2\pi |\omega'|^2 t / \text{Im}(\tau)}$. Now $\lambda_t(z)$ converges absolutely for all $z \in \mathbf{C}$ because

$$\sum_{\omega \in \Omega \smallsetminus \{0\}} \frac{1}{|\omega'|^2} e^{-\frac{2\pi}{\text{Im}(\tau)} |\omega'|^2 t} = S + \sum_{r=1}^{\infty} \sum_{\substack{\omega \in \Omega \\ r \le |\omega'| < r+1}} \frac{1}{|\omega'|^2} e^{-\frac{2\pi}{\text{Im}(\tau)} |\omega'|^2 t}$$

$$\le S + \sum_{r=1}^{\infty} \frac{1}{r^2} e^{-\frac{2\pi}{\text{Im}(\tau)} r^2 t} \#\{\omega \in \Omega;\ |\omega'| \le r\}$$

$$\le S + C \sum_{r=0}^{\infty} e^{-\frac{2\pi}{\text{Im}(\tau)} r^2 t} < +\infty$$

where we used the fact that $\mathbf{Z} + \tau \mathbf{Z}$ contains Cr^2 elements of modulus at most r for some constant $C = C(\tau) > 0$. The equality in (1.30) is to be understood pointwise, even if $z = 0$.

In order to prove Proposition 1.26 we must relate λ to λ_t as t approaches 0 from the right. We begin by showing that λ_t converges to λ in $L^2([0,1)^2)$.

Lemma 1.27 *We have* $\lim_{m \to +\infty} \|\lambda_{1/m} - \lambda\|_{L^2} = 0$.

Proof We can easily compute the norm of $\lambda_{1/m} - \lambda$ using the fact that the functions ψ_ω are pairwise orthogonal and of norm 1 in $L^2([0, 1)^2)$. We get

$$\frac{(2\pi)^2}{\text{Im}(\tau)^2} \|\lambda_{1/m} - \lambda\|_{L^2}^2 = \sum_{\omega \in \Omega \smallsetminus \{0\}} \frac{1}{|\omega'|^4} \left(1 - e^{-\frac{2\pi|\omega'|^2}{\text{Im}(\tau)m}}\right)^2$$

$$= \sum_{\omega \in \Omega \smallsetminus \{0\}} \frac{1}{|\omega|^4} \left(1 - e^{-\frac{2\pi|\omega|^2}{\text{Im}(\tau)m}}\right)^2$$

and will now show that the right-hand side converges to 0 as $m \to +\infty$. We split the sum up into sub-sums over $|\omega|^2 \le m\text{Im}(\tau)/(2\pi)$ and $|\omega|^2 > m\text{Im}(\tau)/(2\pi)$. We bound terms subject to the first condition by $(2\pi/(m\text{Im}(\tau)))^2$ using $1 - e^{-t} \le t$ for all $t > 0$. We bound the other terms by $1/|\omega|^4$. So

$$\frac{(2\pi)^2}{\text{Im}(\tau)^2} \|\lambda_{1/m} - \lambda\|_{L^2}^2 \le \sum_{|\omega|^2 \le m\text{Im}(\tau)/(2\pi)} \frac{4\pi^2}{m^2\text{Im}(\tau)^2} + \sum_{|\omega|^2 > m\text{Im}(\tau)/(2\pi)} \frac{1}{|\omega|^4}.$$

There are at most Cm terms in the first sum where $C > 0$ depends only on τ. Thus

$$\frac{(2\pi)^2}{\text{Im}(\tau)^2} \|\lambda_{1/m} - \lambda\|_{L^2}^2 \le \frac{C}{m} + \sum_{|\omega|^2 > m\text{Im}(\tau)/(2\pi)} \frac{1}{|\omega|^4}$$

after possibly increasing C. The latter sum tends to zero as $m \to +\infty$ since $\sum_{\omega \in \Omega \smallsetminus \{0\}} |\omega|^{-4}$ converges. $\qquad\square$

We now show a variant of Proposition 1.26 for λ_t instead of λ.

Lemma 1.28 *If $t > 0$ and $z_1, \ldots, z_m \in \mathbf{C}$, then*

$$\sum_{1 \le i,j \le m} \lambda_t(z_i - z_j) \ge 0. \qquad (1.31)$$

Proof The proof relies on two basic properties of λ_t. First, the series (1.30) is absolutely convergent and second, its Fourier coefficients are non-negative. We may change the order of summation and obtain

$$\sum_{1 \le i,j \le m} \lambda_t(z_i - z_j) = \frac{\text{Im}(\tau)}{2\pi} \sum_{\omega \in \Omega \smallsetminus \{0\}} \frac{1}{|\omega'|^2} e^{-\frac{2\pi}{\text{Im}(\tau)}|\omega'|^2 t} \sum_{1 \le i,j \le m} e^{2\pi\sqrt{-1}\langle\omega, z_i - z_j\rangle}$$

$$= \frac{\text{Im}(\tau)}{2\pi} \sum_{\omega \in \Omega \smallsetminus \{0\}} \frac{1}{|\omega'|^2} e^{-\frac{2\pi}{\text{Im}(\tau)}|\omega'|^2 t} \left|\sum_{i=1}^{m} e^{2\pi\sqrt{-1}\langle\omega, z_i\rangle}\right|^2$$

$$(1.32)$$

where the second equality used $\langle \omega, z_i \rangle \in \mathbf{R}$. The lemma follows as the right-hand side is clearly non-negative. $\qquad\square$

We observe that the sum in (1.31) allows terms with $i = j$ even though they are forbidden in the context of Proposition 1.26 as λ is not defined at 0.

The sum inside the absolute value in (1.32) is called a Weyl sum and plays an important role in questions of equidistribution.

Let us study the derivative

$$\frac{\partial \lambda_t}{\partial t}(z) = - \sum_{\omega \in \Omega \smallsetminus \{0\}} e^{-\frac{2\pi}{\mathrm{Im}(\tau)}|\omega'|^2 t} \psi_\omega(z) = -g_t(z) + 1 \qquad (1.33)$$

for fixed z, where

$$g_t(z) = \sum_{\omega \in \Omega} e^{-\frac{2\pi}{\mathrm{Im}(\tau)}|\omega'|^2 t} \psi_\omega(z)$$

also sums over $0 \in \Omega$.

We will relate $g_t(z)$ to the classical theta function

$$\vartheta(z; \tau) = \sum_{n \in \mathbf{Z}} e^{\pi\sqrt{-1}n^2\tau + 2\pi\sqrt{-1}nz}$$

defined for $z \in \mathbf{C}$ and $\tau \in \mathbf{H}$. The function $\vartheta : \mathbf{C} \times \mathbf{H} \to \mathbf{C}$ is holomorphic and satisfies two transformation laws

$$\vartheta(z; \tau) = \vartheta(z + 1/2; \tau + 1),$$

$$\vartheta(z; \tau) = (-\sqrt{-1}\tau)^{-1/2} e^{-\frac{\pi}{\tau}\sqrt{-1}z^2} \vartheta(z/\tau; -1/\tau). \qquad (1.34)$$

The first equation is a consequence of the simple fact $n^2 + n \equiv 0 \mod 2$ for all $n \in \mathbf{Z}$. The second one is deeper, see Section 1.2, Chapter 2 [9]. As $\tau \in \mathbf{H}$ we find that $-\sqrt{-1}\tau$ lies in the right-half plane $\{z \in \mathbf{C}; \ \mathrm{Re}(z) > 0\}$. On this domain, we have a holomorphic function that extends the usual square-root function on the positive reals $(0, \infty) \to (0, \infty)$. This is the square-root used above.

From our hypothesis that E has j-invariant in $[1728, +\infty)$ we infer that

$$\mathrm{Re}(\tau) = 0.$$

This property is used in the next lemma to show that g_t is a product of theta functions. It would be interesting to see a generalization of this method to all real j-invariants.

Lemma 1.29 *If* $\mathrm{Re}(\tau) = 0$, *then*

$$g_t(z) = \frac{1}{2t} \sum_{a,b \in \mathbf{Z}} e^{-\frac{\pi}{2t}\left(\frac{(a-u)^2}{\mathrm{Im}(\tau)} + (b-v)^2 \mathrm{Im}(\tau)\right)} > 0 \qquad (1.35)$$

for all $z = u + v\tau \in \mathbf{C}$ with $u, v \in \mathbf{R}$ and all $t > 0$. Moreover, if $z \notin \Omega$, there exists $t_0 \in (0, 1)$ and $C > 0$ such that $g_t(z') \leq C$ for all $t \in (0, t_0]$ and all z' in some neighborhood of $z \in \mathbf{C}$.

Proof We write $\tau = \sqrt{-1}y$ with $y > 0$. By periodicity we may assume without loss of generality that $u, v \in [0, 1)$.

The period lattice Ω is $\mathbf{Z} + \sqrt{-1}y\mathbf{Z}$ and thus a cartesian product. This is crucial for the factorization of

$$g_t(z) = \sum_{a,b \in \mathbf{Z}} e^{-\frac{2\pi}{y}(b^2 + a^2 y^2)t + 2\pi\sqrt{-1}(au+bv)}$$

$$= \left(\sum_{a \in \mathbf{Z}} e^{-2\pi a^2 yt + 2\pi\sqrt{-1}au} \right) \left(\sum_{b \in \mathbf{Z}} e^{-\frac{2\pi}{y}b^2 t + 2\pi\sqrt{-1}bv} \right).$$

So

$$g_t(z) = \vartheta(u; 2ty\sqrt{-1})\vartheta(v; 2ty^{-1}\sqrt{-1}).$$

We transform each factor separately using (1.34) and find that $g_t(z)$ equals

$$(2ty)^{-1/2} e^{-\frac{\pi}{2ty}u^2} \vartheta\left(-\frac{u}{2ty}\sqrt{-1}; \frac{1}{2ty}\sqrt{-1}\right).$$

$$(2t/y)^{-1/2} e^{-\frac{\pi y}{2t}v^2} \vartheta\left(-\frac{vy}{2t}\sqrt{-1}; \frac{y}{2t}\sqrt{-1}\right)$$

$$= \frac{1}{2t} e^{-\frac{\pi}{2t}(u^2/y + v^2 y)} \vartheta\left(-\frac{u}{2ty}\sqrt{-1}; \frac{1}{2ty}\sqrt{-1}\right) \vartheta\left(-\frac{vy}{2t}\sqrt{-1}; \frac{y}{2t}\sqrt{-1}\right).$$

We reinsert the definition of ϑ to find

$$g_t(z) = \frac{1}{2t} e^{-\frac{\pi}{2t}\left(\frac{u^2}{y} + v^2 y\right)} \sum_{a \in \mathbf{Z}} e^{\frac{\pi}{2t}\frac{-a^2 + 2au}{y}} \sum_{b \in \mathbf{Z}} e^{\frac{\pi}{2t}(-b^2 + 2bv)y}$$

$$= \frac{1}{2t} \sum_{a \in \mathbf{Z}} e^{-\frac{\pi}{2t}\frac{(a-u)^2}{y}} \sum_{b \in \mathbf{Z}} e^{-\frac{\pi}{2t}(b-v)^2 y}.$$

This is the equality in (1.35). The inequality follows as all terms are positive.

To prove the upper bound let us assume $u \notin \mathbf{Z}$, the other case $v \notin \mathbf{Z}$ is very similar. We fix a sufficiently small open neighborhood U of u in $\mathbf{R}$ that does not meet $\mathbf{Z}$. The infimum $\inf_{u' \in U} \inf_{a \in \mathbf{Z}} |a - u'|$ is positive. So there is $t_0 \in (0, 1)$ such that if $0 < t < t_0$ then $t' \geq e^{-\pi(a-u')^2/(4y)}$ for all $a \in \mathbf{Z}$ and all $u' \in U$. This inequality entails

$$t^{-1} e^{-\pi\frac{(a-u')^2}{2ty}} \leq e^{-\pi\frac{(a-u')^2}{4ty}}$$

and hence

$$g_t(z) \leq \frac{1}{2} \sum_{a \in \mathbf{Z}} e^{-\pi \frac{(a-u')^2}{4ty}} \sum_{b \in \mathbf{Z}} e^{-\pi \frac{(b-v)^2 y}{2t}} \leq \frac{1}{2} \sum_{a \in \mathbf{Z}} e^{-\pi \frac{(a-u')^2}{4t_0 y}} \sum_{b \in \mathbf{Z}} e^{-\pi \frac{(b-v)^2 y}{2t_0}}$$

if $z = u' + vy\sqrt{-1}$ as $t < t_0$. There is a constant $c > 0$, depending only on U, with $|a - u'| \geq c|a|$ for all $a \in \mathbf{Z}$. Therefore,

$$g_t(z) \leq \frac{1}{2} \sum_{a \in \mathbf{Z}} e^{-\pi \frac{c^2 a^2}{4t_0 y}} \sum_{b \in \mathbf{Z}} e^{-\pi \frac{(b-v)^2 y}{2t_0}}$$

for all $z \in U + \tau v$. The second claim follows as $\sum_{b \in \mathbf{Z}} e^{-\pi \frac{(b-v)^2 y}{2t_0}}$ is continuous in v and thus bounded from above in an open neighborhood of v. $\qquad \square$

The previous lemma and (1.33) imply that the function

$$t \mapsto \lambda_t(z) - t$$

is real analytic and decreasing on $(0, +\infty)$ for fixed $z \in \mathbf{C}$, i.e.

$$\lambda_{t'}(z) - t' \leq \lambda_t(z) - t \tag{1.36}$$

if $0 < t \leq t'$. In particular, it has a limit, possibly $+\infty$, as $t \to 0$ from the right.

Now we prove pointwise convergence of $\lambda_t(z)$ as t approaches 0.

Lemma 1.30 *Suppose $z \in \mathbf{C} \setminus \Omega$. Then $\lim_{m \to +\infty} \lambda_{1/m}(z)$ is a real number and equals $\lambda(z)$.*

Proof Suppose $z \notin \Omega$ and let t_0 be as in Lemma 1.29. If $0 < t \leq t' \leq t_0$ then

$$\lambda_t(z) - t = \int_t^{t'} g_s(z) ds + \lambda_{t'}(z) - t' \leq C(t' - t) + \lambda_{t'}(z) - t' \tag{1.37}$$

with C as in the last lemma. So the proposed limit is a real number $\lambda_0(z) < +\infty$. This yields the existence of the limit in the current lemma.

The upper bound from Lemma 1.29 is uniform over an open neighborhood of z. Combining this with (1.36) and (1.37) we find that

$$|\lambda_t(z') - \lambda_{t'}(z')| \leq (C+1)|t - t'|$$

holds for z' in this neighborhood and for all t, t' with $0 < t \leq t' \leq t_0$. Therefore, λ_t converges locally uniformly to λ_0 on $\mathbf{C} \setminus \Omega$. It follows that $\lambda_0 : \mathbf{C} \setminus \Omega \to \mathbf{R}$ is continuous as λ_t is continuous for all $t > 0$.

The Ω-periodic function λ_1 is continuous. So it is bounded from below by $-M + 1 \leq 0$, say. Let $t' = 1$, inequality (1.36) shows that λ_t is bounded from below by $-M$. We may thus apply Fatou's Lemma, cf. 12.23, Chapter III [11], to the sequence $(\lambda_{1/m} + M)_{m \geq 1}$ where the members are taken as functions on $[0, 1)^2$ in coordinates u, v. We obtain the inequality in

$$\int_{[0,1)^2} (\lambda_0(z) + M) du dv = \int_{[0,1)^2} \liminf_{m \geq 1} (\lambda_{1/m}(z) + M) du dv$$

$$\leq \liminf_{m \geq 1} \int_{[0,1)^2} (\lambda_{1/m}(z) + M) du dv$$

where the value of λ_0 at $z = 0$ is irrelevant. Note $|\lambda_0(z)| \leq \lambda_0(z) + 2M$ on $\mathbf{C} \smallsetminus \Omega$. So

$$\|\lambda_0\|_{L^1} = \int_{[0,1)^2} |\lambda_0(z)| du dv \leq \liminf_{m \geq 1} \int_{[0,1)^2} (|\lambda_{1/m}(z)| + 2M) du dv$$

$$= \liminf_{m \geq 1} \|\lambda_{1/m}\|_{L^1} + 2M \tag{1.38}$$

where $\| \cdot \|_{L^1}$ denotes the L^1-norm on $[0, 1)^2$.

By the Cauchy-Schwarz inequality the L^2-norm dominates the L^1-norm on $[0, 1)^2$; so we have $\|\lambda_{1/m}\|_{L^1} \leq \|\lambda_{1/m}\|_{L^2}$ for all $m \in \mathbf{N}$. By Lemma 1.27, $(\lambda_{1/m})_m$ converges in $L^2([0, 1)^2)$. So this sequence is bounded and this entails $\|\lambda_0\|_{L^1} < +\infty$ after recalling (1.38).

If $z \notin \Omega$, then (1.36) yields $\lambda_{1/m}(z) \leq 1/m + \lambda_0(z)$ and $\lambda_{1/m}(z) \geq \lambda_1(z) - 1$. Therefore,

$$|\lambda_{1/m}(z)| \leq 1 + |\lambda_0(z)| + |\lambda_1(z)|$$

and thus the $\lambda_{1/m}$ are dominated by a measurable function of finite L^1-norm. We may apply the Dominated Convergence Theorem, cf. the proof of 12.30, Chapter III [11], to conclude that $\lambda_{1/m} \to \lambda_0$ in $L^1([0, 1)^2)$ as $m \to +\infty$. We know from Lemma 1.27 that $\lambda_{1/m}$ converges to λ in $L^2([0, 1)^2)$. So it converges to λ in $L^1([0, 1)^2)$ by the Cauchy-Schwarz inequality. A sequence in $L^1([0, 1)^2)$ can have at most one limit point, so $\lambda = \lambda_0$.

Of course, this is an equality of elements in $L^1([0, 1)^2)$. So the functions $\lambda, \lambda_0 : \mathbf{C} \smallsetminus \Omega \to \mathbf{R}$ are equal on the complement of a measure zero set. However, λ_0 is continuous on $\mathbf{C} \smallsetminus \Omega$ by arguments given above and λ is continuous on the same space by general properties of the local Néron-Tate height stated in Theorem 1.22. Therefore, $\lambda(z) = \lambda_0(z)$ for all $z \in \mathbf{C} \smallsetminus \Omega$. $\square$

Proof of Proposition 1.26   It suffices to prove

$$S = \sum_{\substack{1 \leq i,j \leq n \\ i \neq j}} \lambda(z_i - z_j) \geq -\frac{n}{2} \log n - Cn$$

where $z_1, \ldots, z_n$ are elements of $\mathbf{C}$ that are pair-wise distinct modulo Ω and where $C > 0$ depends only on E.

From the preceding lemma we deduce

$$\lambda(z_i - z_j) \geq \lambda_{1/n}(z_i - z_j) - \frac{1}{n}$$

if $i \neq j$ on taking $t' = 1/n$ and $t \to 0$ in (1.36). So

$$S \geq \left(\sum_{\substack{1 \leq i,j \leq n \\ i \neq j}} \lambda_{1/n}(z_i - z_j) \right) - \frac{n(n-1)}{n}$$

$$= \left(\sum_{1 \leq i,j \leq n} \lambda_{1/n}(z_i - z_j) \right) - n\lambda_{1/n}(0) - n + 1;$$

observe that the second sum runs over the diagonal entries $i = j$. The second sum is non-negative by Lemma 1.28 and thus

$$S \geq -n\lambda_{1/n}(0) - n. \tag{1.39}$$

It remains to bound $\lambda_{1/n}(0)$ from above. Baker and Petsche do this explicitly for general elliptic curves in Lemma A.6 [2]. We give another argument using the special expression for g_t given by Lemma 1.29. With $y = \mathrm{Im}(\tau)$ we obtain

$$\lambda_{1/n}(0) - \lambda_1(0) < \lambda_{1/n}(0) - \lambda_1(0) + 1 - \frac{1}{n}$$

$$= \int_{1/n}^1 g_t(0)dt$$

$$= \frac{1}{2} \int_{1/n}^1 \frac{1}{t} \sum_{(a,b) \in \mathbf{Z}^2} e^{-\frac{\pi}{2t}(a^2/y + b^2 y)} dt$$

$$= \frac{1}{2} \int_{1/n}^1 \frac{dt}{t} + \frac{1}{2} \int_{1/n}^1 \frac{1}{t} \sum_{(a,b) \in \mathbf{Z}^2 \setminus \{0\}} e^{-\frac{\pi}{2t}(a^2/y + b^2 y)} dt \tag{1.40}$$

$$= \frac{\log n}{2} + \frac{1}{2} \int_{1/n}^1 \sum_{(a,b) \in \mathbf{Z}^2 \setminus \{0\}} \frac{1}{t} e^{-\frac{\pi}{2t}(a^2/y + b^2 y)} dt.$$

Given $c > 0$, the positive function $t \mapsto e^{-c/t}/t$ tends to 0 as $t \to +\infty$ and $t \to 0$ from the right. Its derivative has a unique zero at $t = c$ on $(0, +\infty)$. Therefore, $e^{-c/t}/t \leq e^{-1}/c$ for all $t > 0$ and $e^{-c/t}/t \leq e^{-c}$ if $t \leq 1 < c$. We apply this with $c = \pi(a^2/y + b^2 y)/2$ to bound terms in (1.40) and conclude that $\lambda_{1/n}(0) - \lambda_1(0)$ is at most

$$\frac{\log n}{2} + \frac{1}{2}\int_{1/n}^{1}\left(\sum_{0<a^2/y+b^2y\leq 2/\pi}\frac{2}{\pi e(a^2/y+b^2y)} + \sum_{a^2/y+b^2y>2/\pi}e^{-\pi(a^2/y+b^2y)/2}\right)dt$$

$$\leq \frac{\log n}{2} + \sum_{0<a^2/y+b^2y\leq 2/\pi}\frac{1}{\pi e(a^2/y+b^2y)} + \frac{1}{2}\sum_{a^2/y+b^2y>2/\pi}e^{-\pi(a^2/y+b^2y)/2}.$$

Only the final sum is over infinitely many terms, but it is clearly convergent. Therefore, $\lambda_{1/n}(0) \leq (\log n)/2 + C$ for some constant C which depends only on Ω. We insert this bound into (1.39) and obtain $S \geq -(n\log n)/2 - Cn - n$. The proof follows after replacing C by $C + 1$. $\qquad\square$

References

[1] J. Ax, *Some topics in differential algebraic geometry I: Analytic subgroups of algebraic groups*, Amer. J. Math. **94** (1972), 1195–1204.

[2] M. Baker and C. Petsche, *Global discrepancy and small points on Elliptic Curves*, Int. Math. Res. Not. (2005), no. 61, 3791–3834.

[3] F.A. Bogomolov, *Points of finite order on abelian varieties*, Izv. Akad. Nauk SSSR Ser. Mat. **44** (1980), no. 4, 782–804, 973.

[4] F.A. Bogomolov, *Sur l'algébricité des représentations l-adiques*, C. R. Acad. Sci. Paris Sér. A-B **290** (1980), no. 15, A701–A703.

[5] E. Bombieri and W. Gubler, *Heights in Diophantine Geometry*, Cambridge University Press, 2006.

[6] W.D. Brownawell and K.K. Kubota, *The algebraic independence of Weierstrass functions and some related numbers*, Acta Arith. **33** (1977), no. 2, 111–149.

[7] J.W.S. Cassels, *Lectures on Elliptic Curves*, London Mathematical Society Student Texts, vol. 24, Cambridge University Press, Cambridge, 1991.

[8] S. David, *Points de petite hauteur sur les courbes elliptiques*, J. Number Theory **64** (1997), no. 1, 104–129.

[9] H.M. Farkas and I. Kra, *Theta Constants, Riemann Surfaces and the Modular Group*, Graduate Studies in Mathematics, vol. 37, American Mathematical Society, Providence, RI, 2001.

[10] G.H. Hardy and E.M. Wright, *An Introduction to the Theory of Numbers*, sixth ed., Oxford University Press, Oxford, 2008.

[11] E. Hewitt and K. Stromberg, *Real and Abstract Analysis. A Modern Treatment of the Theory of Functions of a Real Variable*, Springer-Verlag, New York, 1965.

[12] M. Hindry, *Autour d'une conjecture de Serge Lang*, Invent. Math. **94** (1988), no. 3, 575–603.

[13] M. Hindry and J. Silverman, *Sur le nombre de points de torsion rationnels sur une courbe elliptique*, C. R. Acad. Sci. Paris Sér. I Math. **329** (1999), no. 2, 97–100.

[14] M. Hindry and J.H. Silverman, *Diophantine Geometry An Introduction*, Springer, 2000.

[15] P. Hriljac, *Splitting Fields of Principal Homogeneous Spaces*, Number theory (New York, 1984–1985), Lecture Notes in Math., vol. 1240, Springer, Berlin, 1987, pp. 214–229.

[16] E. Hrushovski, *The Manin-Mumford conjecture and the model theory of difference fields*, Ann. Pure Appl. Logic **112** (2001), no. 1, 43–115.

[17] S. Lang, *Division points on curves*, Ann. Mat. Pura Appl. (4) **70** (1965), 229–234.

[18] ———, *Introduction to Arakelov Theory*, Springer-Verlag, New York, 1988.

[19] T. Loher and D. Masser, *Uniformly counting points of bounded height*, Acta Arith. **111** (2004), no. 3, 277–297.

[20] H.B. Mann, *On linear relations between roots of unity*, Mathematika **12** (1965), 107–117.

[21] D.W. Masser, *Counting points of small height on elliptic curves*, Bull. Soc. Math. France **117** (1989), no. 2, 247–265.

[22] B. Mazur, *Modular curves and the Eisenstein ideal*, Inst. Hautes Études Sci. Publ. Math. (1977), no. 47, 33–186.

[23] L. Merel, *Bornes pour la torsion des courbes elliptiques sur les corps de nombres*, Invent. Math. **124** (1996), no. 1-3, 437–449.

[24] Y. Peterzil and S. Starchenko, *Uniform definability of the Weierstrass ℘ functions and generalized tori of dimension one*, Selecta Math. (N.S.) **10** (2004), no. 4, 525–550.

[25] J. Pila, *O-minimality and the André-Oort conjecture for* $\mathbb{C}^n$, Ann. of Math. (2011), no. 173, 1779–1840.

[26] J. Pila and A.J. Wilkie, *The rational points of a definable set*, Duke Math. J. **133** (2006), no. 3, 591–616.

[27] J. Pila and U. Zannier, *Rational points in periodic analytic sets and the Manin-Mumford conjecture*, Atti Accad. Naz. Lincei Cl. Sci. Fis. Mat. Natur. Rend. Lincei (9) Mat. Appl. **19** (2008), no. 2, 149–162.

[28] R. Pink and D. Roessler, *On Hrushovski's proof of the Manin-Mumford conjecture*, Proceedings of the International Congress of Mathematicians, Vol. I (Beijing, 2002), Higher Ed. Press, Beijing, 2002, pp. 539–546.

[29] ———, *On ψ-invariant subvarieties of semiabelian varieties and the Manin-Mumford conjecture*, J. Algebraic Geom. **13** (2004), no. 4, 771–798.

[30] N. Ratazzi and E. Ullmo, *Galois + équidistribution = Manin-Mumford*, Arithmetic geometry, Clay Math. Proc., vol. 8, Amer. Math. Soc., Providence, RI, 2009, pp. 419–430.

[31] M. Raynaud, *Courbes sur une variété abélienne et points de torsion*, Invent. Math. **71** (1983), no. 1, 207–233.

[32] ———, *Sous-variétés d'une variété abélienne et points de torsion*, Arithmetic and geometry, Vol. I, Progr. Math., vol. 35, Birkhäuser Boston, Boston, MA, 1983, pp. 327–352.

[33] J.-P. Serre, *Propriétés galoisiennes des points d'ordre fini des courbes elliptiques*, Invent. Math. **15** (1972), no. 4, 259–331.

[34] J.-P. Serre, *A Course in Arithmetic*, Springer-Verlag, New York-Heidelberg, 1973, Translated from the French, Graduate Texts in Mathematics, No. 7.

[35] A. Silverberg, *Torsion points on abelian varieties of CM-type*, Compositio Math. **68** (1988), no. 3, 241–249.

[36] J.H. Silverman, *The Arithmetic of Elliptic Curves*, Springer, 1986.

[37] ———, *Advanced Topics in the Arithmetic of Elliptic Curves*, Graduate Texts in Mathematics, vol. 151, Springer-Verlag, New York, 1994.

2

Rational points on definable sets

A. J. Wilkie

Abstract

Most of the papers in this volume depend on what has become known as the o-minimal point counting theorem. The aim of this article is to provide enough background in both model theory and number theory for a graduate student in one, but not necessarily both, of these disciplines to be able to understand the statement and the proof of the theorem.

The one dimensional case of the theorem is treated here in full and differs from the original paper ([PW]) in both its number theoretic side (I use the Thue-Siegel Lemma rather than the Bombieri-Pila determinant method) and in its model theoretic side (where the reparametrization of definable functions is made very explicit). The Thue-Siegel method extends easily to the higher dimensional case but, just as in [PW], one has to revert to Yomdin's original inductive argument to extend the reparametrization, and this is only sketched here.

I am extremely grateful to Adam Gutter for typing up my handwritten notes of the Manchester LMS course on which this paper is based. Also, my deepest thanks to Margaret Thomas for so carefully reading the original manuscript. Her numerous suggestions have greatly improved the presentation.

1 Introduction

So, the aim of these notes is to prove the following:

O-Minimality and Diophantine Geometry, ed. G. O. Jones and A. J. Wilkie. Published by Cambridge University Press. © Cambridge University Press 2015.

1.1 Theorem (Pila-Wilkie [PW])

Let $S \subseteq \mathbb{R}^n$ be a set <u>definable</u> in some <u>o-minimal expansion</u> of the ordered field of real numbers. Assume that S contains no infinite, <u>semi-algebraic</u> subset. Let $\epsilon > 0$ be given. Then for all sufficiently large H, the set S contains at most H^ϵ rational points of <u>height</u> at most H.

• The underlined terms will be defined below.

1.2

For $q \in \mathbb{Q}$, say $q = a/b$ in lowest terms, the <u>height</u> of q, denoted $\mathrm{ht}(q)$, is defined as $\max\{|a|, |b|\}$. For $\overline{q} = \langle q_1, \ldots, q_n \rangle \in \overline{\mathbb{Q}^n}, \mathrm{ht}(\overline{q}) := \max\{\mathrm{ht}(q_1), \ldots, \mathrm{ht}(q_n)\}$.

1.3

A set $S \subseteq \mathbb{R}^n$ is called <u>basic semi-algebraic</u> if it is of the form $\{\overline{a} \in \mathbb{R}^n : P(\overline{a}) > 0\}$ for some polynomial $P(\overline{x}) \in \mathbb{R}[\overline{x}]$.

The collection $\mathcal{A}_n$ of all <u>semi-algebraic</u> subsets of $\mathbb{R}^n$ is defined inductively as follows:

(1) every basic semi-algebraic subset of $\mathbb{R}^n$ is in $\mathcal{A}_n$;
(2) if $X \in \mathcal{A}_n$, then $\mathbb{R}^n \setminus X \in \mathcal{A}_n$;
(3) if $X, Y \in \mathcal{A}_n$, then $X \cup Y \in \mathcal{A}_n$ and $X \cap Y \in \mathcal{A}_n$;
(4) nothing else is in $\mathcal{A}_n$.

1.4 Exercises

(1) Let $P(\overline{x}) \in \mathbb{R}[\overline{x}]$ (where $\overline{x} = (x_1, \ldots, x_n)$). Prove that $Z(P) \in \mathcal{A}_n$, where $Z(P) := \{\overline{a} \in \mathbb{R}^n : P(\overline{a}) = 0\}$.
(2) Suppose that $X \in \mathcal{A}_n$ and $Y \in \mathcal{A}_m$. Prove that $X \times Y \in \mathcal{A}_{n+m}$.
(3) Find an example of a polynomial $P(x) \in \mathbb{R}[x]$ (in the single variable x) such that the closure of the set $\{x \in \mathbb{R} : P(x) > 0\}$ is not the set $\{x \in \mathbb{R} : P(x) \geq 0\}$. Show, however, that (for your example) the closure of $\{x \in \mathbb{R} : P(x) > 0\}$ <u>is</u> a semi-algebraic subset of $\mathbb{R}$.

2 Some semi-algebraic geometry

In fact, it is the case that the closure of any semi-algebraic set is semi-algebraic, but this is very difficult to prove directly. Instead we appeal to the fundamental result of the subject.

2.1 Theorem (Tarski-Seidenberg, see e.g. [D2])

Let $Y \in \mathcal{A}_{n+m}$ and let $X := \pi_n^{n+m}[Y]$ be the projection of Y onto the first n coordinates, i.e.

$$X = \{\bar{x} \in \mathbb{R}^n : \exists \bar{y} \in \mathbb{R}^m \langle \bar{x}, \bar{y} \rangle \in Y\}.$$

Then $X \in \mathcal{A}_n$.

2.2 Exercise

Let $Y \in \mathcal{A}_{n+m}$ and consider the set $X' := \{\bar{x} \in \mathbb{R}^n : \forall \bar{y} \in \mathbb{R}^m \langle \bar{x}, \bar{y} \rangle \in Y\}$. Prove that $X' \in \mathcal{A}_n$.

Now, for any set $X \subseteq \mathbb{R}^n$, observe that the closure $\overline{X}$ of X in $\mathbb{R}^n$ satisfies, for all $\bar{x} \in \mathbb{R}^n$,

$$\bar{x} \in \overline{X} \iff \forall \epsilon \in \mathbb{R}(\epsilon > 0 \implies \exists \bar{y} \in \mathbb{R}^n (\|\bar{x} - \bar{y}\|^2 < \epsilon \text{ and } \bar{y} \in X))$$

(where $\|\cdot\|_n$ denotes the sup norm on $\mathbb{R}^n$).

So if $X \in \mathcal{A}_n$, the expression on the right-hand side here provides a recipe for showing (via uses of 2.1, 2.2 and of the rules 1.3(1), (2), and (3)) that $\overline{X} \in \mathcal{A}_n$.

2.3 Exercises

(1) Complete the proof that $\overline{X} \in \mathcal{A}_n$ whenever $X \in \mathcal{A}_n$. Use a similar method to show that the interior X° of X is in $\mathcal{A}_n$ whenever X is.

(2) Convince yourself that if $f : \mathbb{R}^n \to \mathbb{R}$ is a semi-algebraic function (meaning that its graph is a semi-algebraic subset of $\mathbb{R}^{n+1}$), then the set $\{\bar{x} \in \mathbb{R}^n : f \text{ is continuous at } \bar{x}\}$ is a semi-algebraic subset of $\mathbb{R}^n$.

There are many structure theorems for semi-algebraic sets. For example, every $X \in \mathcal{A}_n$ has the form $X = \bigcup_{i=1}^{N} X_i$, where each X_i is in $\mathcal{A}_n$ and is connected. This is fairly clear for $n = 1$, and it turns out that one can actually deduce the general case from this using little more than the properties 1.3(1), (2), (3), and 2.1. This suggests an axiomatic treatment.

3 O-minimal structures ([PS], [D1])

Let A be any non-empty set and suppose we are given, for each $n \geq 1$, a collection $\mathcal{S}_n$ of subsets of A^n. We write $\mathcal{S}$ for the disjoint union $\bigcup_{n \geq 1} \mathcal{S}_n$. We call $\mathcal{S}$ a structure (on A).

3.1 Definition

The definable closure, $\tilde{\mathcal{S}} = \dot{\bigcup}_{n \geq 1} \tilde{\mathcal{S}}_n$, of $\mathcal{S}$ is the collection of sets defined inductively as follows:

(1) $\mathcal{S}_n \subseteq \tilde{\mathcal{S}}_n$ (and each $X \in \tilde{\mathcal{S}}_n$ is a subset of A^n);

(2) $\{a\} \in \tilde{\mathcal{S}}_1$ for each $a \in A$;

(3) for each i, j with $1 \leq i, j \leq n$, $\{\langle a_1, \ldots, a_n \rangle \in A^n : a_i = a_j\} \in \tilde{\mathcal{S}}_n$;

(4) if $X \in \tilde{\mathcal{S}}_n$ and $Y \in \tilde{\mathcal{S}}_m$ then $X \times Y \in \tilde{\mathcal{S}}_{n+m}$;

(5) if $X, Y \in \tilde{\mathcal{S}}_n$, then $X \cup Y$, $X \cap Y$ and $A^n \setminus X$ are all in $\tilde{\mathcal{S}}_n$;

(6) if $X \in \tilde{\mathcal{S}}_{n+m}$, then $\pi_n^{n+m}[X] \in \tilde{\mathcal{S}}_n$;

(7) nothing else is in $\tilde{\mathcal{S}}$.

Sets in $\tilde{\mathcal{S}}$ are called definable (from, or in, $\mathcal{S}$). Functions whose graphs are in $\tilde{\mathcal{S}}$ are called definable functions.

3.2 Examples and exercises

(1) Say $A = \mathbb{C}$ and $\mathcal{S}_n = \{Z(P) : P(z_1, \ldots, z_n) \in \mathbb{C}[z_1, \ldots, z_n]\}$. By quoting a theorem of Chevalley, prove that $\tilde{\mathcal{S}}_n$ consists precisely of the constructible sets (i.e. $\tilde{\mathcal{S}}_n$ is just the Boolean closure of $\mathcal{S}_n$).

(2) Say $A = \mathbb{N}$ and $\mathcal{S}_n = \{Z(P) \cap \mathbb{N}^n : P(x_1, \ldots, x_n) \in \mathbb{Z}[x_1, \ldots, x_n]\}$. By quoting the MRDP (Matijasevic-Robinson-Davis-Putnam) theorem, prove that $\tilde{\mathcal{S}}_1$ contains a non-computable set.

(3) For arbitrary $A, \mathcal{S}$, and $X \in \tilde{\mathcal{S}}_n$, prove that $X_\sigma \in \tilde{\mathcal{S}}_n$, where σ is any permutation of $\{1, \ldots, n\}$ and $X_\sigma := \{\langle x_{\sigma(1)}, \ldots, x_{\sigma(n)} \rangle \in A^n : \langle x_1, \ldots, x_n \rangle \in X\}$.

Henceforth, we shall assume that $A = \mathbb{R}$ (thereby abandoning many model-theoretic methods) and consider only those structures $\mathcal{S}$ on $\mathbb{R}$ such that for all n, $\mathcal{A}_n \subseteq \mathcal{S}_n$ (i.e. $\mathcal{S}_n$ contains every semi-algebraic subset of $\mathbb{R}^n$). This is what is meant by saying that $\mathcal{S}$ is an expansion of the ordered field of real numbers. Thus, all terms in the statement of our main theorem have now been explained once we give the following

3.3 Definition

A structure $\mathcal{S}$ (on $\mathbb{R}$) is called *o-minimal* if every set $X \in \tilde{\mathcal{S}}_1$ is a finite union of open intervals and singleton sets.

3.4 Remarks and examples

(1) It is crucial in 3.3 that the condition on X (which is equivalent to saying that X has finite boundary) holds for all X in $\tilde{\mathcal{S}}_1$ and not just those X in $\mathcal{S}_1$.

(2) We have seen that the collection $\mathcal{A}$ $(= \tilde{\mathcal{A}})$ of semi-algebraic sets is an o-minimal structure, so at least one exists.

(3) Another example (see [Wi]) is $\mathcal{A}^{\exp}$, where

$$\mathcal{A}_n^{\exp} := \{Z(F) : F(\bar{x}) = P(\bar{x}, e^{\bar{x}}) \text{ for some } P(\bar{x}, \bar{y}) \in \mathbb{R}[\bar{x}, \bar{y}]\}.$$

(We have written $e^{\bar{x}}$ for $(e^{x_1}, \dots, e^{x_n})$ here.) The definable sets here are precisely the projections of sets in $\mathcal{A}^{\exp}$.

(4) In [M], Miller shows that if $\mathcal{S}$ is o-minimal and some function $f : \mathbb{R} \to \mathbb{R}$ of greater than polynomial growth (at ∞) is definable (from $\mathcal{S}$), then $\mathcal{A}^{\exp} \subseteq \tilde{\mathcal{S}}$ (so, of course, $\tilde{\mathcal{A}}^{\exp} \subseteq \tilde{\mathcal{S}}$).

(5) Another example (Denef-van den Dries [DD], Gabrielov [G]) is $\mathcal{A}^{\mathrm{an}}$. Here we take $\mathcal{A}_n^{\mathrm{an}}$ to be the union of $\mathcal{A}_n$ and the collection of all bounded subanalytic subsets of $\mathbb{R}^n$. I won't define this collection here, but suffice it to say that if U is an open subset of $\mathbb{R}^n$ and $f : U \to \mathbb{R}$ is a real analytic function (i.e. it is infinitely differentiable, and for each $\bar{a} \in U$, the Taylor series of f at $\bar{a}$ converges to $f(\bar{x})$ for each $\bar{x} \in \mathbb{R}^n$ sufficiently close to $\bar{a}$), then $f \restriction_K$ is definable in $\mathcal{A}^{\mathrm{an}}$ for each closed box $K \subseteq U$. (As an exercise, exhibit an analytic function $f : (0, 1) \to \mathbb{R}$ which is not definable in any o-minimal structure.)

(6) The largest o-minimal structure required for application in this volume is $\mathcal{A}^{\mathrm{an, exp}} := \mathcal{A}^{\mathrm{an}} \cup \mathcal{A}^{\exp}$. The o-minimality here is due to van den Dries and Miller (see [DM]).

(7) There is no largest o-minimal structure. Indeed, in [RSW] it is shown that if $f : [0, 1] \to \mathbb{R}$ is any infinitely differentiable function, then there exist o-minimal structures $\mathcal{S}_1$ and $\mathcal{S}_2$ and functions $f_i : [0, 1] \to \mathbb{R}$ definable in $\mathcal{S}_i$ (for $i = 1, 2$), such that $f = f_1 + f_2$. (Why does this imply that there is no largest o-minimal structure?)

(8) (Bombieri-Pila, see [BP]) Theorem 1.1 is the best result possible in the sense that the "ϵ" cannot be replaced by any function $\epsilon(H) > 0$ which tends monotonically to 0 as H tends to infinity. This is already the case for the o-minimal structure $\mathcal{A}^{\mathrm{an}}$. However, for the structure $\mathcal{A}^{\exp}$, I conjecture that "H^{ϵ}" may be replaced by $(\log H)^c$ for some constant c (depending on the set S). This is known for definable curves and some surfaces (see Jones-Thomas [JT], Butler [B]).

4 Some 1-dimensional o-minimal theory

Throughout this section, $\mathcal{S}$ is an arbitrary o-minimal structure (on $\mathbb{R}$). For those not used to working "inside" o-minimal structures, the following exercise provides good practice as well as containing an important result.

4.1 Exercise

Let $a < b$ and suppose that $f : (a, b) \to \mathbb{R}$ is a definable function (in $\mathcal{S}$). Prove that $\lim_{x \to b^-} f(x)$ exists (as an element of $\mathbb{R} \cup \{\pm\infty\}$).

The first major result in the subject is:

4.2 The Monotonicity Theorem

Let $f : \mathbb{R} \to \mathbb{R}$ be a definable function. Then there exist real numbers $a_1 < a_2 < \cdots < a_N$ such that (setting $a_0 = -\infty$ and $a_{N+1} = +\infty$) for each $i = 0, \ldots, N, f \upharpoonright_{(a_i, a_{i+1})}$ is continuous and either strictly monotonic or constant.

4.3

For those o-minimal structures that we are considering here (i.e. those with underlying set $\mathbb{R}$), the basic theory (such as the Monotonicity Theorem and its generalization to higher dimensions) was worked out by van den Dries in [D1]. However, the foundations of the full, general theory are due to Pillay and Steinhorn (see [PS]). They discovered (and made precise) the remarkable fact that when a finiteness theorem has been established, then it holds <u>uniformly</u>. As an example of this phenomenon, suppose that $F : \mathbb{R}^{n+m} \to \mathbb{R}^k$ is a definable map. This gives rise to a family

$$\mathcal{F}_F := \{F(\bar{x}, \cdot) : \bar{x} \in \mathbb{R}^n\}$$

of definable maps from $\mathbb{R}^m$ to $\mathbb{R}^k$ *parametrized by* $\mathbb{R}^n$. Such a family is called a <u>definable family of maps</u>.

Now if, in 4.2, $f \in \mathcal{F}_F$, then the N may be chosen to depend only on F, i.e. one may take the same N for all $f \in \mathcal{F}_F$. (We are in the case $m = k = 1$ here.) Further, for each $i = 1, \ldots, N$, the correspondence $f \mapsto a_i$ may be chosen to be definable in the sense that if $f(\cdot) = F(\bar{x}, \cdot)$, then the $a_1, \ldots, a_N$ can be chosen to be definable functions of the parameter $\bar{x}$.

Similarly, if $S \in \mathcal{S}_{n+m}$ then the collection

$$\{\{\bar{y} \in \mathbb{R}^m : \langle \bar{x}, \bar{y} \rangle \in S\} : \bar{x} \in \mathbb{R}^n\}$$

is called a definable family of subsets of $\mathbb{R}^m$. One consequence of the results in [PS] is that for any definable family $\mathcal{B}$ of subsets of $\mathbb{R}^m$ there exists N such that each set in $\mathcal{B}$ is the union of at most N topologically connected sets. Notice that in the case $m = 1$ this is (apparently) stronger than 3.3: the definition merely states (or, rather, is clearly equivalent to the fact) that the number of connected components is finite, with no such uniform bound being required to exist.

4.4 Exercises

(1) Prove this uniformity result directly (for $m = 1$) in the semi-algebraic case (i.e. in the case $\mathcal{S} = \mathcal{A}$).

(2) Let $f : \mathbb{R} \to \mathbb{R}$ be definable. Prove that the set

$$C'_f = \{x \in \mathbb{R} : f \text{ is continuously differentiable on}$$

$$\text{some open neighbourhood of } x\}$$

is definable (and uniformly so in definable families of functions). Deduce that C'_f is cofinite (and hence uniformly so). [Hint: a monotone function (defined on an open interval) is differentiable everywhere except on a set of Lebesgue measure 0.]

Show further that the derivative of f (on its domain C'_f) is definable.

5 Reparametrization (one variable case)

Theorem 1.1 is proved by first reducing to the case that $S \subseteq (0, 1)^n$ (which is easy upon observing that the four functions $x \mapsto \pm x^{\pm 1}$ preserve height (and definability)), and then to the case that S is the image of some definable map $F : (0, 1)^m \to (0, 1)^n$ for some $m < n$. (We assume that $n \geq 2$. Theorem 1.1 is trivial for $n = 1$. Why?)

The idea now is to reparametrize F. This means that we look for a finite set, Φ say, of definable maps $\phi : (0, 1)^m \to (0, 1)^m$ such that $\bigcup_{\phi \in \Phi} \mathrm{Im}(\phi) = (0, 1)^m$ and such that all the maps ϕ and $F \circ \phi$ exhibit more regular behaviour than did F. Indeed, we require that for arbitrary positive integers p (where Φ will depend on p), all these maps be p-times continuously differentiable with all derivatives up to order p being bounded by 1 in absolute value.

In this section, I consider the case $m = 1$.

So let $F : (0, 1) \to (0, 1)^n, x \mapsto \langle F_1(x), \ldots, F_n(x) \rangle$ be a definable map. (In fact, everything must be uniform in definable families of such maps, but we suppress the parameters and simply observe that this is the case.) We assume that one of the F_i's is the identity function (simply by increasing n and adding it as a coordinate function if necessary). By 4.2 and 4.4.(2), there exist $a_0 = 0 < a_1 < \cdots < a_N < 1 = a_{N+1}$ such that each F_i is C^1 on each interval (a_j, a_{j+1}). Further, by considering the definable sets $\{x \in \mathbb{R} : |F_i'(x)| < |F_k'(x)|\}$ (for $1 \leq i, k \leq n$) and using the definition of o-minimality (3.3) we may suppose (possibly after further subdivision) that, for each $j = 0, \ldots, N$ and each $i, k = 1, \ldots, n$, $|F_i'(x)| - |F_k'(x)|$ has constant sign (positive, negative, or zero) throughout (a_j, a_{j+1}).

We first find the set Φ, as described above, for the case $p = 1$.

Fix some j with $1 \leq j \leq N$. Let $x_0 \in (a_j, a_{j+1})$ and choose i_j so that

$$|F_{i_j}'(x_0)| \geq |F_i'(x_0)| \quad \text{for } i = 1, \ldots, n.$$

Then for all $x \in (a_j, a_{j+1})$ we have

$$|F_{i_j}'(x)| \geq |F_i'(x)| \quad \text{for } i = 1, \ldots, n, \text{ and so also} \tag{5.1}$$

$$|F_{i_j}'(x)| \geq 1. \tag{5.2}$$

In particular, F_{i_j} maps (a_j, a_{j+1}) strictly monotonically onto some interval (c, d), where $0 \leq c < d \leq 1$. Now define $\phi_j : (0, 1) \to (0, 1)$ by $x \mapsto F_{i_j}^{-1}(c + (d - c)x)$. Clearly ϕ_j is definable and satisfies, for all $x \in (0, 1)$,

$$F_{i_j}(\phi_j(x)) = c + (d - c)x \tag{5.3}$$

$$\text{and } \operatorname{Im}(\phi_j) = (a_j, a_{j+1}). \tag{5.4}$$

Then, for $x \in (0, 1)$, $|\phi_j'(x)| = \left| \frac{d-c}{F_{i_j}'(\phi_j(x))} \right| \leq |d - c| \leq 1$ (by (5.2), (5.4)).

Further, for $i = 1, \ldots, n$,

$$\begin{aligned}
|(F_i \circ \phi_j)'(x)| &= |F_i'(\phi_j(x))| \cdot |\phi_j'(x)| \\
&= \frac{|F_i'(\phi_j(x))| \cdot |d - c|}{|F_{i_j}'(\phi_j(x))|} \\
&\leq |d - c| \quad \text{(by (5.1), (5.3))} \\
&\leq 1.
\end{aligned}$$

So, taking Φ to be $\{\phi_0, \ldots, \phi_N\}$ together with the constant functions $x \mapsto a_j$ (for $j = 1, \ldots, N$), we have established the following:

5.1 Lemma (the C^1-1-reparameterization lemma)

Let $F = \langle F_1, \ldots, F_n \rangle : (0, 1) \to (0, 1)^n$ be any definable map. Then there exists a finite set Φ of definable functions mapping $(0, 1)$ to $(0, 1)$ having the following two properties. Firstly, both ϕ and $F_i \circ \phi$ are continuously differentiable with derivatives bounded in modulus by 1, for each $\phi \in \Phi$ and $i = 1, \ldots, n$. And secondly, $\bigcup_{\phi \in \Phi} \text{Im}(\phi) = (0, 1)$.

We now aim to improve C^1 to C^p in 5.1. We use the following:

5.2 Lemma

Let $p \geq 1$ and let $I = (a, b)$ be a bounded open interval in $\mathbb{R}$. Suppose that $f : I \to (0, 1)$ is *any* (not necessarily definable) C^{p+1} function having the property that for all $x \in I$ and all $j = 0, \ldots, p + 1$, $f^{(j)}(x) \neq 0$. Then for $j = 0, \ldots, p$ and all $x \in I$,

$$|f^{(j)}(x)| \leq \left(\frac{j+1}{\delta_I(x)}\right)^j,$$

where $\delta_I(x) := \min\{x - a, b - x\}$.

Proof
By linear rescaling it is sufficient to consider $I = (0, 1)$ and to prove that, for all $x \in (0, \frac{1}{2}]$,

$$|f^{(j)}(x)| \leq \left(\frac{j+1}{x}\right)^j$$

for $j = 0, \ldots, p$.

We use induction on j (for all f satisfying the hypotheses). The case $j = 0$ being clear, assume that the lemma holds for some j with $0 \leq j < p$. Since neither $f^{(j+1)}$ nor $f^{(j+2)}$ has a zero (note that $j + 2 \leq p + 1$), it follows that $|f^{(j+1)}|$ is monotonic on $(0, 1)$. Assume first that $|f^{(j+1)}|$ is decreasing on $(0, 1)$. Let $x \in (0, \frac{1}{2}]$. Define $x_0 := \frac{j+1}{j+2} \cdot x$, so that $0 < x_0 < x \leq \frac{1}{2}$. By the Mean Value Theorem, there is some $\xi \in [x_0, x]$ such that

$$f^{(j)}(x) - f^{(j)}(x_0) = f^{(j+1)}(\xi) \cdot (x - x_0). \qquad (1)$$

Since $f^{(j)}$ has no zeros, it follows that $|f^{(j)}(x) - f^{(j)}(x_0)|$ is at most $\max\{|f^{(j)}(x)|, |f^{(j)}(x_0)|\}$ which, by the inductive hypothesis, is bounded by $\left(\frac{j+1}{x_0}\right)^j = \left(\frac{j+2}{x}\right)^j$. Further, since $|f^{(j+1)}|$ is decreasing, $|f^{(j+1)}(\xi)| \geq |f^{(j+1)}(x)|$. Also, $|x - x_0| = \frac{x}{j+2}$. Putting the last three remarks into equation (1), we obtain $\left(\frac{j+2}{x}\right)^{j+1} \geq |f^{(j+1)}(x)|$, as required.

Now if $|f^{(j+1)}|$ is increasing on $(0, 1)$, we consider the function $g : (0, 1) \to (0, 1), x \mapsto f(1 - x)$. This satisfies all the hypotheses of the theorem, and

hence too the inductive hypothesis: $|g^{(j)}(x)| \le \left(\frac{j+1}{x}\right)^j$ (for all $x \in (0, \frac{1}{2}]$).
But $|g^{(j+1)}(x)|$ is decreasing on $(0,1)$, so we may apply the argument above to
obtain $|g^{(j+1)}(x)| \le \left(\frac{j+2}{x}\right)^{j+1}$ (for all $x \in (0, \frac{1}{2}]$).

Now for $x \in (0, \frac{1}{2}]$, $x \le 1 - x$, so $|f^{(j+1)}(x)| \le |f^{(j+1)}(1 - x)|$ (as $|f^{(j+1)}|$
is increasing on $(0,1)$). But $|f^{(j+1)}(1 - x)| = |g^{(j+1)}(x)|$, so we obtain that
$|f^{(j+1)}(x)| \le \left(\frac{j+2}{x}\right)^{j+1}$ in this case too. $\qquad\square$

5.3 Exercises

(1) Formulate and prove a many variable version of 5.2. (Unfortunately, this does not seem to help in proving the many variable reparametrization lemma.)
(2) Suppose that $f : (0,1) \to (0,1)$ satisfies the hypotheses of 5.2 for *all p*. Prove that the series

$$U(z) = U(x + \sqrt{-1}y) := \sum_{j=0}^{\infty} \frac{f^{(j)}(x)}{j!} \cdot (\sqrt{-1}y)^j$$

converges absolutely and uniformly on any compact subset of the region $\{x + \sqrt{-1}y : e \cdot |y| < x \le \frac{1}{2}\}$. Deduce that f has a complex analytic continuation to this region and hence that f itself, which was only assumed to be a C^∞ function, is in fact real analytic. (An old theorem of Bernstein (see, for example, [KP]) asserts that if $f : (0,1) \to (0,1)$ is a C^∞ function with $f^{(j)}(x) > 0$ for all $j = 0,1,2,\ldots$ and all $x \in (0,1)$, then f is real analytic.)

5.4 Lemma

Let $p \ge 1$ and let I be any bounded open interval in $\mathbb{R}$. Suppose that $f : I \to (0,1)$ is any C^{p+1} function having the property that for all $x \in I$ and all $j = 0,\ldots,p+1$, $f^{(j)}(x) \ne 0$. Assume further that $|f'(x)| \le 1$ for all $x \in I$. Then for $j = 1,\ldots,p$ and all $x \in I$ we have

$$|f^{(j)}(x)| \le \left(\frac{j}{\delta_I(x)}\right)^{j-1}.$$

Proof
    Apply 5.2 with f' in place of f. $\qquad\square$

The previous lemmas suggest a reparametrization of the form $f((\delta_I(x))^p)$. So we need a formula for the higher derivatives of composite functions.

5.5 Exercise

Let $f : I \to \mathbb{R}, g : J \to I$ be any C^p functions, where I, J are open subsets of $\mathbb{R}$. Show that for any $x \in J$ and $q = 1, \ldots, p$, $(f \circ g)^{(q)}(x)$ has the form

$$\sum_{k=1}^{q} f^{(k)}(g(x)) \cdot \left[\sum_{v=1}^{(q,k)} B_q(k_1, \ldots, k_q) \cdot \prod_{v=1}^{q} (g^{(v)}(x))^{k_v} \right]$$

for some positive integers $B_q(k_1, \ldots, k_q)$ (independent of f, g) and where the inner summation is over all q-tuples $\langle k_1, \ldots, k_q \rangle$ of non-negative integers satisfying the conditions

$$k_1 + k_2 + \cdots + k_q = k \quad \text{and} \quad k_1 + 2k_2 + \cdots + qk_q = q.$$

[Hint: Clearly the derivative of the composite function has the given form for a summation over *some* finite set of q-tuples $\langle k_1, \ldots, k_q \rangle$, with the coefficients $B_q(k_1, \ldots, k_q)$ independent of f, g. So one may use one's favourite functions $(x^\alpha, e^x, \ldots)$ to discover which $B_q(k_1, \ldots, k_q)$ are 0.]

5.6 Remark

The above formula is attributed to the Blessed Francesco Faà di Bruno (1825–1888) (see [KP]). The values of the coefficients are given by

$$B_q(k_1, \ldots, k_q) = \frac{q!}{\prod_{v=1}^{q} k_v!(v!)^{k_v}}$$

(where $0! = 1$).

5.7 Theorem (the C^p-1-reparametrization theorem)

Let $F : (0, 1) \to (0, 1)^n$ be any definable map. Then for any $p \geq 1$, there exists a finite set Φ of definable functions mapping $(0, 1)$ to $(0, 1)$ such that $\bigcup_{\phi \in \Phi} \text{Im}(\phi) = (0, 1)$ and such that for each $\phi \in \Phi$, $i = 1, \ldots, n$, and $q = 0, \ldots, p$, both $\phi^{(q)}$ and $(F_i \circ \phi)^{(q)}$ exist, are continuous, and are bounded in modulus by 1 on $(0, 1)$, where F_i is the ith coordinate function of F. (Further, $|\Phi|$ depends only on p and uniformly on F, as do the functions in Φ.)

Proof

It is easily seen that it is sufficient to prove the theorem with the bound 1 in the conclusion replaced with some function of p (independent of F). Further we may assume, by first applying 5.1 and then replacing F by a tuple of all the

ϕ's and $F_i \circ \phi$'s so obtained (which entails a rather large increase in n), that F is a C^1 function with the first derivatives of its coordinate functions bounded by 1.

Now, by repeated use of 4.4(2) and the usual subdivision method, there exist $a_0 = 0 < a_1 < \cdots < a_N < a_{N+1} = 1$ such that F is a C^{p+1} map on each interval (a_j, a_{j+1}) and for each $v = 1, \ldots, p+1$, each coordinate function of $F^{(v)}$ is either identically zero or has no zeros on (a_j, a_{j+1}). Define $\phi_j : (0,1) \to (a_j, a_{j+1})$ by $\phi_j(x) := a_j + \frac{1}{2}(a_{j+1} - a_j)x^p$. Then $|\phi_j^{(q)}(x)| \le p!$ for $x \in (0,1)$ and for each $q = 0, \ldots, p$. Also, for $\langle k_1, \ldots, k_q \rangle$ a q-tuple as in 5.5, we have

$$\left| \prod_{v=1}^{q} (\phi_j^{(v)}(x))^{k_v} \right| = \left| \prod_{v=1}^{q} ((p(p-1)\cdots(p-v+1)) \cdot \left(\frac{a_{j+1} - a_j}{2} \right) \cdot x^{p-v})^{k_v} \right|$$

$$\le \prod_{v=1}^{q} p^{v \cdot k_v} \cdot \left(\frac{a_{j+1} - a_j}{2} \right)^{k_v} \cdot x^{pk_v - vk_v}$$

$$= p^q \cdot \left(\frac{a_{j+1} - a_j}{2} \right)^{k} \cdot x^{pk-q} \qquad \text{(for } x \in (0,1)\text{)}.$$

We now apply 5.5 and 5.2 with $I = (a_j, a_{j+1})$, $J = (0,1)$, $g = \phi_j$, and $f = F_i \restriction_{(a_j,a_{j+1})}$, where F_i is some coordinate function of F.

Note that ϕ_j actually maps into (and in fact, onto) the left open half of the interval I, so that $\delta_I(\phi_j(x)) = \frac{1}{2}(a_{j+1} - a_j)x^p$ for all $x \in (0,1)$. So it follows from 5.4 that for all $x \in (0,1)$ and all $k = 1, \ldots, q$, $|f^{(k)}(\phi_j(x))| \le \left(\frac{2k}{(a_{j+1}-a_j)x^p} \right)^{k-1}$. Applying 5.5, we obtain (for certain integers $B_q'(k)$, $B''(p)$ depending only on p)

$$|(f \circ \phi_j)^{(q)}(x)| \le \sum_{k=1}^{q} \left(\frac{2k}{(a_{j+1} - a_j)x^p} \right)^{k-1} \cdot p^q \cdot x^{pk-q} \cdot \left(\frac{a_{j+1} - a_j}{2} \right)^{k} \cdot B_q'(k)$$

$$= x^{p-q} \cdot p^q \cdot \left(\frac{a_{j+1} - a_j}{2} \right) \cdot \sum_{k=1}^{q} k^{k-1} \cdot B_q'(k)$$

$$\le B''(p) \qquad \text{(since } q \le p\text{)}.$$

[Exercise: Use 5.6 to show that $B''(p)$ may be taken to be $c_1 \cdot p^{c_2 p}$ for some small explicit constants c_1, c_2.]

A similar calculation applies to the function $a_{j+1} - \phi_j$, which maps onto the right open half of the interval I. Thus, the proof of 5.7 is now complete upon taking Φ to be $\{\phi_j, a_{j+1} - \phi_j : j = 0, \ldots, N\}$ together with the constant functions with values $a_1, \ldots, a_N, \frac{a_0 + a_1}{2}, \ldots, \frac{a_N + a_{N+1}}{2}$. $\qquad \square$

5.8 Remark

The parenthetical comment in the statement of 5.7 amounts to this: for F ranging over a definable family of maps $F(\bar{x}, \cdot) : (0,1) \rightarrow (0,1)^n$ (see 4.3) the number N of subintervals in the above proof stays bounded (i.e. has an upper bound independent of $\bar{x}$) and the endpoints a_j are given by N definable functions of the parameters $\bar{x}$. It follows that the reparametrizing functions in Φ depend definably on $\bar{x}$. (Of course, this is all for a fixed $p \geq 1$.)

6 Proof of 1.1 in the 1-dimensional case

There is a well-defined notion of dimension for definable sets which we shall come to later. It turns out that definable subsets of $(0,1)^n$ having dimension 1 are precisely the images of definable maps $f : (0,1) \rightarrow (0,1)^n$. The 1-dimensional case of 1.1 thus reduces to the following:

6.1 Theorem

Let $f : (0,1) \rightarrow (0,1)$ be definable and assume that graph(f) contains no infinite semi-algebraic subset. Then for any $\epsilon > 0$, there exists $C = C_\epsilon(f)$ such that for all $H > C$, there are fewer than H^ϵ pairs $\langle q_1, q_2 \rangle$ of rational numbers of height at most H such that $f(q_1) = q_2$.

6.2 Exercise

Show that graph(f) (f as above) contains no infinite semi-algebraic subset if and only if, for all a, b with $0 < a < b < 1$ and all non-zero polynomials $P(x,y) \in \mathbb{R}[x,y]$, there exists $\alpha \in (a,b)$ such that $P(\alpha, f(\alpha)) \neq 0$.

The proof of 6.1 has three stages. The second of these stages is pure transcendental number theory and requires no o-minimality (nor definability) at all:

6.3 Lemma

Let p and d be integers satisfying $100 \leq 4p \leq d^2 \leq 5p$ and let $\phi, \psi : (0,1) \rightarrow (0,1)$ be C^p functions whose derivatives (of all orders $\leq p$) are bounded by 1 in absolute value.

Then for all $\alpha \in (0,1)$ and all $H > 4 \cdot (2d - 2)^{d/4}$, there exists a non-zero polynomial $P(X_1, X_2) \in \mathbb{Z}[X_1, X_2]$ of degree at most $d - 1$ in each variable such that $P(q_1, q_2) = 0$ for all rationals $q_1, q_2 \in (0,1)$ satisfying

(1) $\mathrm{ht}(\langle q_1, q_2 \rangle) \leq H$ and
(2) there exists $\beta \in (0, 1)$ with $|\beta - \alpha| < H^{-20/d}$ such that $\phi(\beta) = q_1$ and
    $\psi(\beta) = q_2$.

For the proof, we use the following version of the Dirichlet Box Principle (i.e. the Pigeon Hole Principle). I leave the proof as an exercise, or see [W] (page 132, Lemma 4.11).

6.4 Proposition (Thue-Siegel)

Let $v \geq 1, \mu \geq 0$ be integers and for each i, j with $1 \leq i \leq v, 0 \leq j \leq \mu$, let $v_{i,j}$ be a real number. Let u, X, ℓ be positive integers satisfying

(1) $u \geq \max\limits_{0 \leq j \leq \mu} \sum\limits_{i=1}^{v} |v_{i,j}|$ and
(2) $\ell^{\mu+1} < (X + 1)^v$.

Then there exist integers $A_1, \ldots, A_v$ such that

(3) $0 < \max\limits_{1 \leq i \leq v} |A_i| \leq X$, and

(4) $\max\limits_{0 \leq j \leq \mu} \left| \sum\limits_{i=1}^{v} A_i \cdot v_{i,j} \right| \leq \frac{u \cdot X}{\ell}$. $\square$

6.5 Proof of 6.3

Let $\theta_1(x), \ldots, \theta_{d^2}(x)$ be an enumeration of the functions $\phi(x)^s \psi(x)^t$ $(0 \leq s, t < d)$ and consider the function $G : (0, 1) \to \mathbb{R}$ defined by

$$G(x) := \sum_{i=1}^{d^2} A_i \cdot \theta_i(x),$$

where the A_i are integers (to be chosen later) satisfying $|A_i| \leq H^d$ for $i = 1, \ldots, d^2$.

Now the point is that if $\beta \in (0, 1)$ is such that both $\phi(\beta)$ and $\psi(\beta)$ are rationals of height at most H, then either $G(\beta) = 0$ or else $|G(\beta)| \geq \frac{1}{H^{2d-2}}$. (For if $\phi(\beta) = \frac{a_1}{b_1}, \psi(\beta) = \frac{a_2}{b_2}$, then $G(\beta) = \frac{L}{b_1^{d-1} b_2^{d-1}}$ for some $L \in \mathbb{Z}$.) So if we can choose the A_i's so that $|G(\beta)| < \frac{1}{H^{2d-2}}$ for such β that also satisfy $|\beta - \alpha| < H^{-20/d}$, then we are done: just take

$$P(X_1, X_2) = \sum_{\substack{0 \leq s < d \\ 0 \leq t < d}} A_{\lceil \langle s, t \rangle \rceil} X_1^s X_2^t,$$

(where $\lceil \langle s,t \rangle \rceil$ denotes the i such that $\theta_i(x) = \phi(x)^s \psi(x)^t$). In fact, we shall choose the A_i's so that $|G(\beta)| < \frac{1}{H^{2d-2}}$ for *all* $\beta \in (0,1)$ with $|\beta - \alpha| < H^{-20/d}$.

To do this we apply Taylor's Theorem around α:

$$G(x) = \sum_{i=1}^{d^2} A_i \left(\sum_{j=0}^{p-1} \frac{\theta_i^{(j)}(\alpha)}{j!} \cdot (x-\alpha)^j + \frac{\theta_i^{(p)}(\xi_x^i)}{p!}(x-\alpha)^p \right) \qquad (*)$$

for some ξ_x^i's lying between α and x.

6.6 Exercise

Check that $|\theta_i^{(j)}| \le (2d-2)^j$ on $(0,1)$ for $j = 1,\ldots,p$ and for $i = 1,\ldots,d^2$. [Hint: Apply the Leibniz formula for the higher derivatives of a product.]

We first bound the remainder term in $(*)$:

$$\left| \sum_{i=1}^{d^2} A_i \frac{\theta_i^{(p)}(\xi_x^i)}{p!}(x-\alpha)^p \right|.$$

If $|x - \alpha| < H^{-20/d}$, then using 6.6 and the bound $|A_i| \le H^d$, we see that it is at most

$$d^2 \cdot H^d \cdot (2d-2)^p \cdot H^{\frac{-20p}{d}}.$$

Since $4p \le d^2 \le 5p$, this is bounded by

$$d^2 \cdot (2d-2)^{d^2/4} \cdot H^{-3d}$$

$$= \frac{d^2 \cdot (2d-2)^{d^2/4}}{H^{d+2}} \cdot \frac{1}{H^{2d-2}}$$

$$\le \left(\frac{2 \cdot (2d-2)^{d/4}}{H} \right)^d \cdot \frac{1}{H^{2d-2}}$$

$$< \frac{1}{2^d \cdot H^{2d-2}} \qquad \text{(by our hypothesis on } H\text{)}$$

$$< \frac{1}{2 \cdot H^{2d-2}}.$$

We shall be done if we can choose the A_i's so that the main term in $(*)$ has the same bound. By interchanging the order of summation in $(*)$ we see that this main term is

$$\left| \sum_{j=0}^{p-1} (x-\alpha)^j \left(\sum_{i=1}^{d^2} A_i \cdot \frac{\theta_i^{(j)}(\alpha)}{j!} \right) \right|. \qquad (**)$$

This suggests applying 6.4 with $v_{i,j} = \frac{\theta_i^{(j)}(\alpha)}{j!}$. The best values for the other parameters in 6.4 then turn out to be $v = d^2$, $\mu = p - 1$, $\ell = H^{4d}$, $X = H^d$, and $u = d^2 \cdot e^{2d-2}$, which is justified since (by 6.6), in this case

$$\max_{0 \le j \le \mu} \sum_{i=1}^{v} |v_{i,j}| \le \max_{0 \le j \le p-1} \sum_{i=1}^{d^2} \frac{(2d-2)^j}{j!} \le d^2 \cdot e^{2d-2}$$

and

$$\ell^p = H^{4dp} \le H^{d^3} \le (H^d + 1)^{d^2} = (X+1)^v,$$

so 6.4(1), (2) both hold.

So we may indeed find integers A_i (for $1 \le i \le d^2$) satisfying $\max_{1 \le i \le d^2} |A_i| \le H^d$ and

$$\max_{0 \le j \le \mu} \left| \sum_{i=1}^{d^2} A_i \cdot \frac{\theta_i^{(j)}(\alpha)}{j!} \right| \le \frac{u \cdot X}{\ell} = \frac{d^2 \cdot e^{2d-2} \cdot H^d}{H^{4d}} = \frac{d^2 \cdot e^{2d-2}}{H^{3d}}.$$

With this choice of the A_i's, it follows (upon using the trivial estimate $|(x - \alpha)^j| < 1$) that the main term (∗∗) is bounded (for any $x \in (0,1)$) by

$$p \cdot d^2 \cdot \frac{e^{2d-2}}{H^{3d}} \le d^4 \cdot \frac{e^{2d-2}}{H^{3d}}$$

which, by our hypothesis on H and d, is easily seen to be bounded by $\frac{1}{2 \cdot H^{2d-2}}$, as required. □

The first stage in the proof of 6.1 involves reducing the problem to the situation of 6.3. So let $\epsilon > 0$ be given. Let d be an integer such that $d > 10$ and $\frac{20}{d} < \frac{\epsilon}{2}$. Then we may choose an integer p such that $100 \le 4p \le d^2 \le 5p$. Let $f : (0,1) \to (0,1)$ be as in the hypotheses of 6.1. For $H > 4 \cdot (2d-2)^{d/4}$ define

$$S_H := \{\langle q_1, q_2 \rangle \in \mathbb{Q}^2 : f(q_1) = q_2 \text{ and } \mathrm{ht}(\langle q_1, q_2 \rangle) \le H\}$$

and assume, for a contradiction, that $|S_H| \ge H^\epsilon$ for infinitely many H.

Now apply 5.7 (with $n = 1$) and let $c_\epsilon = |\Phi|^{-1}$. [Note that f is now fixed and so d and p, and hence $|\Phi|$, depend only on ϵ. This is also true for f ranging over some fixed definable family of functions.] Then by the Pigeon Hole Principle, there is some fixed $\phi \in \Phi$ such that

$$|S_H \cap \{\langle \phi(\beta), f(\phi(\beta)) \rangle : \beta \in (0,1)\}| \ge c_\epsilon H^\epsilon$$

for infinitely many H.

By subdividing $(0, 1)$ into $1 + \lfloor \frac{H^{20/d}}{2} \rfloor$ intervals of length at most $2 \cdot H^{-20/d}$, it follows, again by the Pigeon Hole Principle, that there is some $\alpha_H \in (0, 1)$ such that, setting

$$Y_H := S_H \cap \{ \langle \phi(\beta), f(\phi(\beta)) \rangle : \beta \in (0, 1) \text{ and } |\alpha_H - \beta| < H^{-20/d} \}$$

we have

$$|Y_H| \geq \frac{c_\epsilon \cdot H^\epsilon}{1 + \left[\frac{H^{20/d}}{2} \right]} \geq c_\epsilon \cdot H^{\epsilon - 20/d} > c_\epsilon \cdot H^{\epsilon/2} \qquad (***)$$

for infinitely many H.

This completes stage one.

We now apply 6.3 (with $\psi = f \circ \phi$) to obtain a polynomial $P_H(X_1, X_2) \in \mathbb{Z}[X_1, X_2]$ of degree at most $d - 1$ in each variable, which vanishes on Y_H.

Now stage three, the final contradiction, involves what is known as "a zero estimate" in transcendental number theory. One requires an upper bound on the number of zeros of the function

$$\sum_{0 \leq s, t < d} A_{\lceil \langle s, t \rangle \rceil} \cdot y^s \cdot f(y)^t$$

which depends only on d (and not on the coefficients). In more concrete situations, one has more information about the function f and good bounds may be found (using complex analysis, say, if f is known to have an entire analytic continuation), giving rise to much sharper bounds in place of H^ϵ in our main theorem here. However, for our purposes we only need to know that *some* such bound exists, N_d say. For then we have our contradiction to the inequality $(***)$ simply by choosing $H > \left(\frac{N_d}{c_\epsilon} \right)^{2/\epsilon}$ (and H satisfying $(***)$).

The existence of N_d follows from our discussion in 4.3. Just consider the definable family

$$\mathcal{F} := \{\{y : \sum_{0 \leq s, t < d} x_{s,t} \cdot y^s \cdot f(y)^t = 0\} : x_{s,t} \in \mathbb{R} \text{ for } 0 \leq s, t < d\}$$

of subsets of $\mathbb{R}$. We choose N_d greater than the number of connected components of any member of $\mathcal{F}$ and observe, via 6.2, that such components are, in fact, singleton sets. □

7 Some remarks on the proof of the general case of 1.1

7.1

One requires some deeper o-minimality theory. In particular, we require a generalization of 5.7 for definable functions $F : (0, 1)^m \to (0, 1)$, and also a result telling us that in order to prove 1.1, it is sufficient to consider the case that S is the graph of such a function.

The first and second stages of the argument discussed in Section 6 may now be generalized in a routine manner: under the assumption that graph(F) contains more than H^ϵ rational $(n + 1)$-tuples of height $\leq H$, we end up with a polynomial $P_H(X_1, \ldots, X_n, X_{n+1})$ (with integer coefficients) of degree depending only on ϵ, such that at least H^ϵ of these points lie in the set $Z(P_H) \cap$ graph(F).

7.2

The third stage, however, requires an inductive argument for which we require a good notion of dimension for definable sets. The proof is then completed as follows.

If, for some sufficiently large H as above, $\dim(Z(P_H) \cap \text{graph}(F)) = \dim(\text{graph}(f)) (= n)$, then it is easily shown that for some sufficiently small open box in $(0, 1)^{n+1}$, Δ_H say, $\dim(\Delta_H \cap Z(P_H)) = n$ and $\Delta_H \cap Z(P_H) = \Delta_H \cap \text{graph}(F)$. Thus, graph($F$) contains the infinite semi-algebraic set $\Delta_H \cap Z(P_H)$, contrary to our assumptions.

Thus, we may assume that there are infinitely many H such that $Z(P_H) \cap$ graph(F) contains at least H^ϵ rational points of height $\leq H$ *and* such that $\dim(Z(P_H) \cap \text{graph}(F)) < \dim(\text{graph}(F))$. It might now appear that, by the obvious inductive argument, we have reached the desired contradiction. However, the inductive hypothesis is being applied to a set, $Z(P_H) \cap \text{graph}(F)$, whose definition depends on H, which is not exactly how Theorem 1.1 is stated. But, as I have been emphasizing throughout, all our arguments have been uniform over definable families and, indeed, all the sets $Z(P_H) \cap \text{graph}(F)$ do lie in one fixed family (depending on ϵ, but not on H). So the correct formulation of 1.1 is as follows:

7.3 A uniform version of Theorem 1.1

Let $\{S_{\bar{x}} : \bar{x} \in \mathbb{R}^k\}$ be a definable family of subsets of $\mathbb{R}^n$. Then for each $\epsilon > 0$ there exists an integer $D_\epsilon > 0$ such that for all $\bar{x} \in \mathbb{R}^k$

either $(1)_{\bar{x}}$ for all $H \geq D_{\epsilon}$, $S_{\bar{x}}$ contains at most H^{ϵ} rational points of height at most H,

    or $(2)_{\bar{x}}$ $S_{\bar{x}}$ contains an infinite semi-algebraic subset.

Further, there exists a definable family $\{A_{\bar{x}} : \bar{x} \in \mathbb{R}^k\}$ of semi-algebraic subsets of $\mathbb{R}^n$ such that for each $\bar{x} \in \mathbb{R}^k$, either $A_{\bar{x}}$ is infinite and is a subset of $S_{\bar{x}}$, or else it is empty (say) and $(1)_{\bar{x}}$ holds.

7.4 Exercise

Check that our proof in Section 6 does in fact establish 7.3 in the case that each $S_{\bar{x}}$ is 1-dimensional.

7.5 The transcendental part of a set

The other papers in this volume require a slightly different formulation of 1.1.

7.6 Definition

For $S \subseteq \mathbb{R}^n$, S^{alg} denotes the union of all infinite, connected, semi-algebraic subsets of S. We also define $S^{\text{trans}} := S \setminus S^{\text{alg}}$.

    Our methods also give the following version of 1.1.

7.7 Theorem

Let $S \subseteq \mathbb{R}^n$ be a set definable in some o-minimal expansion of the ordered field of real numbers. Let $\epsilon > 0$. Then for all sufficiently large H, the set S^{trans} contains at most H^{ϵ} rational points of height at most H.

    In order to complete our discussion of the proof of Theorem 1.1 (and of Theorem 7.7) it now only remains to indicate how one establishes the general case of the reparametrization theorem.

8 Some higher-dimensional o-minimal theory

This section is based on the notions of cell and cell decomposition. Since we may assume (in 1.1 and 7.7) that $S \subseteq (0,1)^n$ (see the comments at the beginning of Section 5), we treat only bounded cells here. All definability is with respect to a fixed o-minimal expansion of the real field.

8.1 Definition

For $n \geq 1$ and $n \geq m \geq 0$ we define the notion of an _m-dimensional cell in $\mathbb{R}^n$_ inductively as follows:

(1) (i) A 0-dimensional cell in $\mathbb{R}$ is a singleton set $\{a\}$ (for $a \in \mathbb{R}$).

    (ii) A 1-dimensional cell in $\mathbb{R}$ is an open interval (a, b) (for $a, b \in \mathbb{R}, a < b$).

(2) For $n \geq 2$, an $(m + 1)$-dimensional cell in $\mathbb{R}^n$ has one of the following forms:

    (i) graph(f), where $f : C \rightarrow \mathbb{R}$ is a definable, bounded, continuous function and C is an $(m + 1)$-dimensional cell in $\mathbb{R}^{n-1}$, or

    (ii) $(f, g)_C := \{\langle \bar{x}, y \rangle \in \mathbb{R}^n : \bar{x} \in C \text{ and } f(\bar{x}) < y < g(\bar{x})\}$, where $f, g : C \rightarrow \mathbb{R}$ are definable, bounded, continuous functions with $f(\bar{x}) < g(\bar{x})$ (for all $\bar{x} \in C$) and C is an m-dimensional cell in $\mathbb{R}^{n-1}$.

8.2 Definition

(1) A finite collection, $\mathcal{C}$ say, of cells in $\mathbb{R}^n$ (of various dimensions) is called a _cell decomposition_ of $(0, 1)^n$ if it partitions $(0, 1)^n$ and, in the case that $n > 1$, the collection $\{\pi_{n-1}^n[C] : C \in \mathcal{C}\}$ is a cell-decomposition of $(0, 1)^{n-1}$.

(2) A cell decomposition $\mathcal{C}$ of $(0, 1)^n$ is called _compatible_ with a set $B \subset (0, 1)^n$ if B is the union of a subcollection of $\mathcal{C}$ (i.e. for all $C \in \mathcal{C}$, either $C \subseteq B$ or $C \cap B = \emptyset$).

The main foundational result of the subject, due to Pillay and Steinhorn (see [PS]), is the following:

8.3 Theorem (the Cell Decomposition theorem for bounded sets)

Suppose that $B_1, \ldots, B_k$ are definable subsets of $(0, 1)^n$. Then there exists a cell decomposition of $(0, 1)^n$ that is compatible with each B_i (for $i = 1, \ldots, k$).

8.4 Definition

The _dimension_ of a definable set $B \subseteq (0, 1)^n$ is defined to be the largest m such that B contains an m-dimensional cell in $\mathbb{R}^n$.

8.5 Exercises

(1) Prove by induction that an m-dimensional cell in $\mathbb{R}^n$ is definably homeomorphic to $(0, 1)^m$.
(2) Prove that an m-dimensional cell in $\mathbb{R}^n$ is open if and only if $m = n$.

8.6

It can be shown that this notion of dimension has good properties with respect to definable sets and definable functions:

$$\dim(A \cup B) = \max\{\dim(A), \dim(B)\}, \qquad \dim(A \times B) = \dim(A) + \dim(B),$$

$$\dim(f[A]) \le \dim(A), \qquad \dim(\bar{A}) = \dim(A), \qquad \dim(\bar{A} \setminus A) < \dim(A)$$

(where we set $\dim(\emptyset) := -1$).

8.7

Further, for any definable map $F : (0, 1)^m \to \mathbb{R}^n$ and $p \ge 0$, there exists a cell decomposition C_p of $(0, 1)^m$ such that F is C^p on each open cell in C_p. In general, this is the best one can do: we cannot replace C^p by C^∞ here. However, in all the o-minimal structures of interest in this volume (in particular, for $\mathcal{A}^{\mathrm{an, exp}}$) one can replace C^p by C^ω (i.e. real analytic).

8.8

There is one further result that we need for the next final section, namely the Principle of Definable Choice. This states that if $S \subseteq \mathbb{R}^{n+m}$ is any definable set, then there exists a definable function $g : \mathbb{R}^n \to \mathbb{R}^m$ such that for all $\bar{x} \in \mathbb{R}^n$, if there exists $\bar{y} \in \mathbb{R}^m$ such that $\langle \bar{x}, \bar{y} \rangle \in S$, then $\langle \bar{x}, g(\bar{x}) \rangle \in S$.

9 Reparametrization (many variable case)

I conclude these notes with a brief sketch of the proof of the following:

9.1 Theorem (the C^p-m-reparametrization theorem)

Let $F : (0, 1)^m \to (0, 1)^n$ be any definable map. Then, for any $p \ge 1$, there exists a finite set Φ of C^p maps mapping $(0, 1)^m$ to $(0, 1)^m$ such that $\bigcup_{\phi \in \Phi} \mathrm{Im}(\phi) = (0, 1)^m$ and such that for each $\phi \in \Phi$ and $\alpha \in \mathbb{N}^m$ with $|\alpha| \le p$, we have that $F \circ \phi$ is C^p and both $\|\phi^{(\alpha)}\|_m$ and $\|(F \circ \phi)^{(\alpha)}\|_n$ are bounded by 1 on $(0, 1)^m$. (Further, $|\Phi|$ depends only on p and uniformly on F, as do the functions in Φ.)

9.2 Remarks

(1) Recall that $\| \cdot \|_k : \mathbb{R}^k \to \mathbb{R}$ denotes the sup norm, $\|\langle x_1, \ldots, x_k \rangle\|_k :=$ $\max\{|x_1|, \ldots, |x_k|\}$, on $\mathbb{R}^k$.

(2) We are using the usual multi-index notation: for $\alpha = \langle \alpha_1, \ldots, \alpha_m \rangle \in \mathbb{N}^m$, $|\alpha| := \alpha_1 + \cdots + \alpha_m$ and, for a $C^{|\alpha|}$ map $f : (0, 1)^m \to (0, 1)^\ell, f^{(\alpha)} :=$ $\frac{\partial^{(|\alpha|)}f}{\partial x_1^{\alpha_1} \cdots \partial x_m^{\alpha_m}}$.

9.3

The reparametrization theorem is due, at least in the semi-algebraic case, to Yomdin and Gromov (see [Gr]). However, the inductive part of the argument, which I now sketch, is due entirely to Yomdin, and once one has the basics of o-minimality in place (Section 8), one has to change very little in generalizing his argument from the semi-algebraic to the o-minimal case.

Firstly, it follows quite easily by induction (using 8.7 and a C^p version of 8.5(1)) that one may assume that F (in 9.1) is already C^p on $(0, 1)^m$. The problem is to bound the derivatives. We may assume that $m \geq 2$ (by 5.7). Let $k \geq 0$.

Now assume that $\Phi = \Phi_k$ has been found to satisfy the conclusion of 9.1 except that this conclusion is weakened from "for all $\alpha \in \mathbb{N}^m$ with $|\alpha| \leq p$" to

"for all $\alpha \in \mathbb{N}^m$ with $|\alpha| \leq p$ and $\alpha_m \leq k$". $(*)_k$

We complete the proof by showing how to construct Φ_{k+1} to satisfy $(*)_{k+1}$. (The case $k = 0$ is dealt with by using a subsidiary induction on m. This involves a use of uniformity (relegating the variable x_m to parameter status) and one must, in fact, incorporate the parenthetical statement at the end of the statement of Theorem 9.1 into the inductive hypothesis. But I do not go into details in this sketch. The interested reader will be in a good position to fill these in after assimilating the inductive step that now follows.)

To this end we consider the map $\widetilde{F} : (0, 1)^m \to (0, 1)^{\tilde{n}}$, where $\tilde{n} = |\Phi_k|(|\Delta| \cdot n + m)$ and where

$$\Delta := \{\alpha = \langle \alpha_1, \ldots, \alpha_m \rangle \in \mathbb{N}^m : |\alpha| \leq p - 1, \alpha_m \leq k\}.$$

The map $\widetilde{F}$ then takes $\bar{x} \in (0, 1)^m$ to an enumeration of the values $(F \circ \phi)^{(\alpha)}((\bar{x}))$ for $\alpha \in \Delta$ and $\phi(\bar{x})$ for $\phi \in \Phi_k$. Notice that $\widetilde{F}$ is a C^1 map and (because of the condition "$|\alpha| \leq p - 1$" in the definition of Δ) that $\|\frac{\partial \tilde{F}}{\partial x_i}(\bar{a})\|_{\tilde{n}} \leq 1$ for each $i = 1, \ldots, m - 1$ and $\bar{a} \in (0, 1)^m$. The required construction of Φ_{k+1} is now easily obtained from the following

Main Lemma

Let $G : (0, 1)^m \to (0, 1)^\ell$ be a definable C^1 map and suppose that $\|\frac{\partial G}{\partial x_i}\|_\ell \leq 1$ on $(0, 1)^m$ for $i = 1, \ldots, m - 1$. Then for any $p \geq 0$, there exists a finite set Φ of C^p functions $\phi : (0, 1) \to (0, 1)$ with $\bigcup_{\phi \in \Phi} \mathrm{Im}(\phi) = (0, 1)$ such that for each $\phi \in \Phi$ and $q = 0, \ldots, p$, $\|\phi^{(q)}\|_1 \leq 1$ and for each $i = 1, \ldots, m$, $\|\frac{\partial}{\partial x_i}(G \circ \tilde{\phi})\|_\ell \leq 1$ on $(0, 1)^m$, where $\tilde{\phi} : (0, 1)^m \to (0, 1)^m : \langle x_1, \ldots, x_m \rangle \mapsto \langle x_1, \ldots, x_{m-1}, \phi(x_m) \rangle$.

Proof sketch

I give the proof in the case that $\frac{\partial G}{\partial x_m}$ is bounded. (The general case follows by considering the restriction of G to the set $(\eta, 1 - \eta)^m$ and then letting $\eta \to 0$. The limiting process is quite routine as the set Φ may be constructed uniformly in η. One also uses 4.1 here.)

So, for each $x_m \in (0, 1)$ we consider a point $\theta(x_m) = \langle \theta_1(x_m), \ldots, \theta_{m-1}(x_m) \rangle \in (0, 1)^{m-1}$ such that

$$\left\| \frac{\partial G}{\partial x_m}(\theta(x_m), x_m) \right\|_\ell \geq \frac{1}{2} \sup \left\{ \left\| \frac{\partial G}{\partial x_m}(x', x_m) \right\|_\ell : x' \in (0, 1)^{m-1} \right\}. \qquad (*)$$

By Definable Choice (8.8) we may suppose that $\theta : (0, 1) \to (0, 1)^{m-1}$ is definable.

Apply 5.7 to the map $H : (0, 1) \to (0, 1)^{m-1+\ell}$ given by $y \mapsto \langle \theta(y), G(\theta(y), y) \rangle$. I claim that the finite set Φ of functions provided by 5.7 also (nearly) works here as well. So let $\phi \in \Phi$. We must show that

$$\left\| \frac{\partial(G \circ \tilde{\phi})}{\partial x_m}(a', a_m) \right\|_\ell \leq 1 \qquad \text{for all } \langle a', a_m \rangle \in (0, 1)^m,$$

the other bounds being straightforward.

Now, the conclusion of 5.7 tells us, in particular, that the derivative of the map $y \mapsto G(\theta(\phi(y)), \phi(y))$ is bounded by 1 on $(0, 1)$, as is the derivative of the map $\theta \circ \phi$. Thus, for each $b \in (0, 1)$,

$$1 \geq \left\| \sum_{i=1}^{m-1} \frac{\partial G}{\partial x_i}(\theta(\phi(b)), \phi(b)) \cdot (\theta_i \circ \phi)'(b) + \frac{\partial G}{\partial x_m}(\theta(\phi(b)), \phi(b)) \cdot \phi'(b) \right\|_\ell$$

$$\geq \left\| \frac{\partial G}{\partial x_m}(\theta(\phi(b)), \phi(b)) \cdot \phi'(b) \right\|_\ell - \left\| \sum_{i=1}^{m-1} \frac{\partial G}{\partial x_i}(\theta(\phi(b)), \phi(b)) \cdot (\theta_i \circ \phi)'(b) \right\|_\ell$$

$$\geq \left\| \frac{\partial G}{\partial x_m}(\theta(\phi(b)), \phi(b)) \cdot \phi'(b) \right\|_\ell - (m-1).$$

(The last inequality follows from the above together with the Lemma hypothesis on $\frac{\partial G}{\partial x_i}$ for $i = 1, \ldots, m-1$.)

Now let $\langle a', a_m \rangle \in (0,1)^m$. Then

$$m \geq \left\| \frac{\partial G}{\partial x_m}(\theta(\phi(a_m)), \phi(a_m)) \cdot \phi'(a_m) \right\|_\ell$$

$$\geq \frac{1}{2} \cdot \left\| \frac{\partial G}{\partial x_m}(a', \phi(a_m)) \cdot \phi'(a_m) \right\|_\ell \qquad \text{(by (*))}$$

$$= \frac{1}{2} \cdot \left\| \frac{\partial (G \circ \tilde\phi)}{\partial x_m}(a', a_m) \right\|_\ell .$$

This gives the bound $2m$ instead of the required 1. However, just as in 5.7 (see the first paragraph of the proof there), this can be easily dealt with by a suitable linear substitution. □

This concludes our discussion of the o-minimal point counting theorem (Theorem 1.1). Further progress would seem to depend on a closer analysis of the proof of reparameterization and in particular on obtaining good bounds for the size of the set Φ of parametrizing maps. I refer the interested reader to the survey [Y] for a discussion of this and related topics.

References

[B]   Butler, Lee A., "Some cases of Wilkie's conjecture", Bull. London Math. Soc., Vol. 44, no. 4, 2012, 642–660.

[BP]   Bombieri, E. and Pila, J., "The number of integral points on arcs and ovals", Duke Math. J., 59, no. 2, 1989, 337–357.

[DD]   Denef, J. and van den Dries, L., "p-adic and real subanalytic sets", Annals of Math. (2nd series) 128, 1988, 80–138.

[D1]   van den Dries, L., "Remarks on Tarski's problem concerning $(\mathbb{R}, +, \cdot, \exp)$", Logic Colloquium 1982, Studies in Logic and the foundations of Mathematics, Vol. 112, 1984, 97–121.

[D2]   van den Dries, L., "Tame topology and o-minimal structures", LMS lecture Notes Series 248, CUP, 1998.

[DM]   van den Dries, L. and Miller, C., "On the real exponential field with restricted analytic functions", Israel J. of Math., 85, 1994, 19–56.

[G]   Gabrielov, A., "Projections of semianalytic sets", Funct. Anal. Appl., 2, 1968, 282–291.

[Gr]   Gromov, M., "Entropy, homology and semialgebraic geometry", Astérisque 145-146, 1987, 225–250.

[JT]   Jones, G. O. and Thomas, M. E. M., "The density of algebraic points on certain Pfaffian surfaces", Q. J. Math., 63, 2012, 637-651.

[KP]   Krantz, S. and Parks, H., "A primer of real analytic functions", 2nd edition, Birkhäuser, 2002.

[M]   Miller, C., "Exponentiation is hard to avoid", Proc. AMS, 122, No. 1, 1994, 257–259.

[PW]   Pila, J. and Wilkie, A. J., "The rational points of a definable set", Duke Math. J., 133, No. 3, 2006, 591–616.

[PS]   Pillay, A. and Steinhorn, C., "Definable sets and ordered structures I", Trans. AMS, 295, 1986, 565–592.

[RSW]   Rolin, J.-P., Speissegger, P. and Wilkie, A. J., "Quasianalytic Denjoy-Carleman classes and o-minimality", J. AMS, 16, 2003, 751–777.

[W]   Waldschmidt, M., "Diophantine approximation on linear algebraic groups", Grund. Math. Wiss., 326, Springer, 2000.

[Wi]   Wilkie, A. J., "Model completeness results for expansions of the ordered field of real numbers by restricted Pfaffian functions and the exponential function", J. AMS, 9, 1996, 1051–1094.

[Y]   Yomdin, Y., "Smooth parameterizations in dynamics, analysis, diophantine and computational geometry", preprint, 2014.

3

Functional transcendence via o-minimality

Jonathan Pila

These notes are an edited version of notes prepared for a series of five lectures delivered during the LMS-EPSRC Short Course on "O-minimality and Diophantine Geometry", held at the University of Manchester, 8-12 July 2013. Not everything here was covered in the lectures, in particular many details were skipped.

Synopsis We describe Schanuel's conjecture, the differential version ("Ax-Schanuel"), and their modular analogues. We sketch proofs of the "Ax-Lindemann" parts in both settings using o-minimality, and describe connections with the Zilber-Pink conjecture.

1 Algebraic independence

Definition 1.1 Let L be a field and $K \subset L$ a subfield.

1. An element $x \in L$ is called *algebraic over K* if there exists a polynomial $p \in K[X]$, non-zero, such that $p(x) = 0$.
2. Elements $x_1, \ldots, x_n \in L$ are called *algebraically dependent over K* if there is a polynomial $p \in K[X_1, \ldots, X_n]$, non-zero, such that $p(x_1, \ldots, x_n) = 0$. Otherwise, they are called *algebraically independent over K*.

Proposition 1.2 *Let L be a field, $K \subset L$ a subfield, and $x \in L$. The following assertions are equivalent:*

1. *x is algebraic over K.*
2. *There exists a finite dimensional K-vector space $V \subset L$ such that $xV \subset V$.*

O-Minimality and Diophantine Geometry, ed. G. O. Jones and A. J. Wilkie. Published by Cambridge University Press. © Cambridge University Press 2015.

Proof **Exercise.** □

Corollary 1.3 *The collection of $x \in L$ which are algebraic over K form a subfield of L (containing K).* **Exercise.** □

An element that is not algebraic (over K) will be called *transcendental over K*. If K is not specified, we take it to be $\mathbb{Q}$; if L is also not specified we take it to be $\mathbb{C}$. We denote by $\overline{\mathbb{Q}}$ the field of algebraic numbers in $\mathbb{C}$, i.e. the elements of $\mathbb{C}$ that are algebraic over $\mathbb{Q}$.

Definition 1.4 Let L be a field and $K \subset L$ a subfield.

1. A *transcendence basis* for L over K is a maximal algebraically independent (over K) subset.

[Transcendence bases exist; if T is a transcendence basis for L over K then every element of L is algebraic over $K(T)$; any two transcendence bases of L over K have the same cardinality. (**Exercises**: Like vector space dimension. Hint: use the

Steinitz Exchange Principle 1.5 Let $K \subset L$ be fields and $u, v, w_1, \dots, w_k$, $y \in L$. Say that v *depends on u over* $w_1, \dots, w_k$ if v is algebraic over $K(w, u)$ but not over $K(w)$. Suppose that v depends on u over $w_1, \dots, w_k$. Then u depends on v over $w_1, \dots, w_k$ and if y is algebraic over $K(w, u)$ then it is algebraic over $K(w, v)$.)]

2. The *transcendence degree* of L over K is the cardinality of a transcendence basis; it is denoted tr.d.$_K L$ or tr.d.(L/K). If S is a set we will also write tr.d.$_K S$ for tr.d.$_K K(S)$, and also tr.d.(S) for tr.d.$_{\mathbb{Q}} S$.

Example 1.6 tr.d.$(\overline{\mathbb{Q}}) = 0$; tr.d.$(\mathbb{C}) = 2^{\aleph_0}$; tr.d.$_{\mathbb{C}}(\mathbb{C}(X_1, \dots, X_n)) = n$, *for independent indeterminates X_i*; tr.d.$_{\mathbb{Q}}(K) = $ tr.d.$_{\overline{\mathbb{Q}}}(K)$.

2 Transcendental numbers

References for this section are Baker [5], Lang [31], Nesterenko [43], Waldschmidt [65]. The existence of transcendental numbers (in $\mathbb{C}$ over $\mathbb{Q}$) may be established by a counting argument: the set of algebraic numbers is countable, but the set of complex numbers is uncountable. Before Cantor's theory of transfinite sets was formulated, Liouville showed in 1844 that certain fast-converging series are transcendental, because irrational algebraic numbers do not admit "very good" approximations by rationals; e.g. $\sum 10^{-n!}$ ([5]).

But the first "naturally occurring" number to be shown transcendental was e, the base of the natural logarithm, by Hermite in 1873. The transcendence of π was proved a little later by Lindemann in 1882. Lindemann stated the following generalisation of his results, a full proof being given by Weierstrass (1885).

Theorem 2.1 (Lindemann/Lindemann-Weierstrass) *Let* $x_1, \ldots, x_n$ *be algebraic numbers which are linearly independent over* $\mathbb{Q}$. *Then*

$$e^{x_1}, \ldots, e^{x_n}$$

are algebraically independent over $\mathbb{Q}$.   □

The transcendence of e follows ($n = 1, x_1 = 1$); the transcendence of any non-zero logarithm of an algebraic number also follows, in particular the transcendence of (πi and hence of) π.

Observe that the result is sharp: if $x_1, \ldots, x_n$ are linearly dependent over $\mathbb{Q}$ then $e^{x_1}, \ldots, e^{x_n}$ are algebraically dependent over $\mathbb{Q}$. (Just exponentiate the linear relation and use the functional property of exp.)

Euler had expressed the view that a ratio of logarithms, if irrational, must be transcendental, e.g. $\log 3/\log 2$. This became Hilbert's 7th problem and was later solved independently in the 30s by Gelfond and Schneider.

The logarithm function is multivalued. One can define principal values which are single valued, but in general for a non-zero complex number a we will use $\log a$ to denote any complex number c with $e^c = a$.

The non-zero elements of a field K will be denoted $K^\times$.

Theorem 2.2 (Gelfond-Schneider) *Let* $a, b \in \overline{\mathbb{Q}}^\times$. *Then* $\log b/\log a$ *is either rational or transcendental.*   □

Otherwise put: if $a \in \overline{\mathbb{Q}}$ is algebraic and not equal to 0 or 1 and $r \in \overline{\mathbb{Q}} - \mathbb{Q}$ then $b = a^r$ is transcendental (else $r = \log b/\log a$ contradicts the theorem). E.g. $2^{\sqrt{2}}$ is transcendental. Also $e^\pi = (-1)^{-i}$. This theorem was generalised by Baker.

Theorem 2.3 (Baker, 1966) *Suppose* $a_1, \ldots, a_n \in \overline{\mathbb{Q}}^\times$ *and* $\log a_1, \ldots,$ $\log a_n$ *are linearly independent over* $\mathbb{Q}$. *Then*

$$1, \log a_1, \ldots, \log a_n$$

are linearly independent over $\overline{\mathbb{Q}}$.   □

The methods of proof of these theorems will not be relevant to our discussion of the *functional* versions, and we won't discuss them.

The functional relation $\exp(x + y) = \exp(x).\exp(y)$ forces e.g. $\log 6, \log 2,$ $\log 3$ to be algebraically dependent. On the general principle that "numbers defined using exponentiation should be as algebraically independent as permitted by the functional relation" one expects things like:

Conjecture on algebraic independence of logarithms 2.4 If $a_1, \ldots, a_n \in \overline{\mathbb{Q}}^{\times}$ are *multiplicatively independent* (i.e. there are no non-trivial multiplicative relations $\prod a_i^{k_i} = 1, k_i \in \mathbb{Z}$) then any determination of the $\log a_i$ is algebraically independent. However, it is not known that $\mathrm{tr.d.}(\log \overline{\mathbb{Q}}) > 1$. Baker's Theorem (above) is the strongest result known here.

Gelfond's Conjecture 2.5 Suppose $a \in \overline{\mathbb{Q}} - \{0, 1\}$ and b is algebraic of degree d. Then the numbers $a^b, a^{b^2}, \ldots, a^{b^{d-1}}$ are algebraically independent; one knows this for $d = 2, 3$ (Gelfond); in general it is known that $\mathrm{tr.d.}(a^b, a^{b^2}, \ldots, a^{b^{d-1}}) \geq \left[\frac{d+1}{2}\right]$ (Brownawell, Waldschmidt, Philippon, Nestrenko, Diaz; see [43]).

Various conjectures 2.6 e^e is transcendental; $e, e^e, \ldots$ are algebraically independent, $e + \pi$ is transcendental and e and π are algebraically independent. A result of Brownawell/ Waldschmidt (1974/3) implies (see Baker): *Either e^e or e^{e^2} is transcendental.* A result of Nesterenko (1996) implies: π *and e^{π} are algebraically independent* (see [43]).

The "four exponentials" conjecture 2.7 Suppose $x_1, x_2 \in \mathbb{C}$ are l.i./$\mathbb{Q}$, and that $y_1, y_2 \in \mathbb{C}$ are l.i./$\mathbb{Q}$. Then at least one of the four exponentials $\exp(x_i y_j)$ is transcendental. E.g. suppose t is irrational (so $1, t$ are l.i./$\mathbb{Q}$). Since $\log 2, \log 3$ are l.i./$\mathbb{Q}$, at least one of $2, 3, 2^t, 3^t$ should be transcendental. This is not known: indeed it is not known that if $t \in \mathbb{R}$ and $2^t, 3^t$ are both integers then $t \in \mathbb{N}$. If one has three x_i then the conclusion is known ("six exponentials"; Siegel, Lang, Ramachandra; see [31]).

3 Schanuel's conjecture

Schanuel, in the 1960s, came up with a conjecture that implies all the theorems and conjectures above, succinctly summarising all the expected transcendence properties of the exponential function. It is stated in Lang's book [31, p30].

Schanuel's Conjecture 3.1 (SC) Let $x_1, \ldots, x_n \in \mathbb{C}$ be linearly independent over $\mathbb{Q}$. Then

$$\mathrm{tr.d.}_{\mathbb{Q}}(x_1, \ldots, x_n, e^{x_1}, \ldots, e^{x_n}) \geq n.$$

For example, when all the x_i are algebraic we recover Lindemann's theorem, and when all the e^{x_i} are algebraic we recover the conjecture on algebraic independence of logarithms.

Exercise   Deduce the other statements above from SC (the ones involving π and e take a bit of work).

References: Kirby [26], Lang [31], Waldschmidt [65, 66], Zilber [71, 72] and especially [73].

4 Differential fields

According to Ax [3], Schanuel made the same conjecture for power series and (more generally) for differential fields (i.e. fields with derivations). As a reference, see Lang [32].

Definition 4.1   A *differential field* is a pair (K, D) where K is a field and $D : K \to K$ is a *derivation*: an additive function satisfying the Leibniz rule: $D(xy) = xDy + yDx$.

In a differential field, the kernel of D is a field (**Exercise**) called the *field of constants*. It always contains the prime field. More generally we will deal with fields with several (commuting) derivations. The derivations D of a field K form a vector space over K: $(zD)(x) = z(Dx)$. Let L be finitely generated over K, of transcendence degree r. Denote by $\mathcal{D}$ the vector space of derivations of L over K (i.e. trivial on K). We have a pairing $(\mathcal{D}, L) \to L$ given by

$$(D, x) \mapsto Dx.$$

Thus each $x \in L$ gives an element dx of the dual space of $\mathcal{D}$, and we have $d(yz) = ydz + zdy, d(y+z) = dy + dz$. These form a subspace of the dual space of $\mathcal{D}$ if we define ydz by $(D, ydz) = yDz$.

Proposition 4.2 (see Lang [32, VIII, 5.5])   *Let L be a separably generated and finitely generated extension of a field K, of transcendence degree r. Then the vector space $\mathcal{D}$ (over L) of derivations of L which are trivial on K has dimension r. Elements $t_1, \ldots, t_r \in L$ form a separating transcendence basis of L over K iff $dt_1, \ldots, dt_r$ form a basis for the dual of $\mathcal{D}$ over L.* $\square$

Example 4.3   *The paradigm examples are fields of functions, where the derivation is induced by differentiation.*

1. In algebraic geometry one considers an (affine) algebraic set $V \subset \mathbb{C}^n$ defined as the locus of common zeros of some set of polynomials in $\mathbb{C}[X_1, \ldots, X_n]$, and hence of the ideal I generated by them. The *coordinate ring*

$$\mathbb{C}[V] = \mathbb{C}[X_1, \ldots, X_n]/I$$

is the ring of functions induced on V by $\mathbb{C}[X_1, \ldots, X_n]$. This ring is a domain just if I is prime, and then V is called irreducible (over $\mathbb{C}$), or an *(affine) variety* (though this word is used with a lot of flexibility), and then it has a quotient field $\mathbb{C}(V)$, which is called an algebraic function field.

Such fields have derivations: on $\mathbb{C}(X)$ one has the derivative with respect to X, which extends non-trivially to $\mathbb{C}(V)$ if X is non-constant on V. If dim $V = k$ it has k independent derivations D_i (over $\mathbb{C}$) given by extending the derivations corresponding to k independent coordinates X_i (as $\mathbb{C}(V)$ is finitely separably generated over $\mathbb{C}$). E.g. if V : $F(X, Y) = 0$ the derivation D with $DX = 1$ extends to $\mathbb{C}(V)$ with $DY = -F_X/F_Y$.

2. The same in an analytic context: let V be a complex analytic variety. Then the field K of meromorphic functions on V has dim V independent (over K, trivial on $\mathbb{C}$) derivations coming from differentiation with respect to suitably chosen coordinate functions (say $V \subset \mathbb{C}^N$).

In both examples, one can add exponentials of any finite number of elements to these fields, perhaps restricting to a neighbourhood of some point of V, in such a way that if $y = \exp(x)$ then $Dy = yDx$ for the derivations mentioned.

5 Ax-Schanuel

We follow Ax's terminology. We consider a tower of fields $\mathbb{Q} \subset C \subset K$ and a set of derivations $\mathcal{D} = \{D_1, \ldots, D_m\}$ on K with $C = \bigcap_j \ker D_j$. By "rank" below we mean rank over K.

Definition 5.1 Elements $x_1, \ldots, x_n \in K$ are called *linearly independent over* $\mathbb{Q}$ *modulo* C, which we write "l.i./$\mathbb{Q}$ mod C", if there is no non-trivial relation

$$\sum_{i=1}^{n} q_i x_i = c, \quad q_i \in \mathbb{Q}, c \in C$$

where non-trivial means not all q_i, c are zero.

Definition 5.2 Elements $y_1, \ldots, y_n \in K^\times$ are called *multiplicatively independent modulo* C if there is no non-trivial relation

$$\prod_{i=1}^{n} y_i^{k_i} = c, \quad k_i \in \mathbb{Z}, c \in C$$

where non-trivial means that not all $k_i = 0$.

Theorem 5.3 ("Ax-Schanuel"; Ax [3], 1971) *Let $x_i, y_i \in K^\times, i = 1, \ldots, n$ with*

(a) $D_j y_i = y_i D_j x_i$ for all j, i,
(b) the x_i are l.i. over $\mathbb{Q}$ modulo C [or (b') the y_i are mult. indpt. over C].

Then

$$\text{tr.d.}_C C(x_1, \ldots, x_n, y_1, \ldots, y_n) \geq n + \text{rank}(D_j x_i)_{i=1,\ldots,n, j=1,\ldots,m}. \quad \square$$

The proof (like the setting) is differential algebra. See also Ax [4], Kirby [25], Brownawell-Kubota, Bertrand-Pillay [9] for generalisations, including to the semi-abelian setting.

We now consider this statement in a complex setting. We take

$$\pi : \mathbb{C}^n \to (\mathbb{C}^\times)^n$$

given by

$$\pi(z_1, \ldots, z_n) = (\exp z_1, \ldots, \exp z_n).$$

Let $A \subset U$ be a complex analytic subvariety of some open set $U \subset \mathbb{C}^n$, so that locally the coordinate functions $z_1, \ldots, z_n$ and $\exp(z_1), \ldots, \exp(z_n)$ are meromorphic on A, and we have derivations $\{D_j\}$ with $\text{rank}(D_j z_i) = \dim A$, the rank being over the field of meromorphic functions, and with $D_j e^{z_i} = e^{z_i} D z_j$ for all i, j. (I.e. we take the D_j to be differentiation with respect to some choice of $\dim A$ independent coordinates on A.)

"Complex Ax-Schanuel" Conjecture 5.4   In the above setting, if the z_i are linearly independent over $\mathbb{Q}$ modulo $\mathbb{C}$, then

$$\text{tr.d.}_{\mathbb{C}} \mathbb{C}(z_1, \ldots, z_n, \exp(z_1), \ldots, \exp(z_n)) \geq n + \dim A.$$

This clearly implies a weaker "two-sorted" version where the transcendence degree of the z_i and $\exp(z_i)$ are computed separately: with the same setting and hypotheses,

$$\text{tr.d.}_{\mathbb{C}} \mathbb{C}(z_1, \ldots, z_n) + \text{tr.d.}_{\mathbb{C}} \mathbb{C}(\exp(z_1), \ldots, \exp(z_n)) \geq n + \dim A.$$

Definition 5.5   A subvariety $W \subset \mathbb{C}^n$ will be called *geodesic* if it is defined by (any number ℓ of) equations of the form

$$\sum_{i=1}^{n} q_{ij} z_i = c_j, \quad j = 1, \ldots, \ell,$$

where $q_{ij} \in \mathbb{Q}, c_j \in \mathbb{C}$.

Definition 5.6 By a *component* we mean a complex-analytically irreducible component of $W \cap \pi^{-1}(V)$ where $W \subset \mathbb{C}^n$ and $V \subset (\mathbb{C}^\times)^n$ are algebraic subvarieties.

Let A be a component of $W \cap \pi^{-1}(V)$. We can consider the coordinate functions z_i and their exponentials as elements of the field of meromorphic functions (at least locally) on A, and we can endow this field with $\dim A$ derivations $\{D_j\}$ as above with $\operatorname{rank}(D z_i) = \dim A$. Then (with Zcl denoting Zariski closure)

$$\dim W \geq \dim \operatorname{Zcl}(A) = \operatorname{tr.d.}_{\mathbb{C}}(z_i), \quad \dim V \geq \dim \operatorname{Zcl}(\exp(A)) = \operatorname{tr.d.}_{\mathbb{C}}(\exp z_i)$$

and the "two-sorted" Ax-Schanuel conclusion becomes

$$\dim W + \dim V \geq \dim X + \dim A$$

provided that the functions z_i are l.i. over $\mathbb{Q} \bmod \mathbb{C}$.

This last condition is equivalent to A not being contained in a proper geodesic subvariety. Let us take U' to be the smallest geodesic subvariety of $\mathbb{C}^n$ containing A. Let $X' = \exp U'$, which is a coset of an algebraic subtorus of $(\mathbb{C}^\times)^n$, and put $W' = W \cap U'$, $V' = V \cap X'$. We can choose coordinates $z_i, i = 1, \ldots, \dim A$ which are l.i. over $\mathbb{Q} \bmod \mathbb{C}$ and derivations as previously with $\operatorname{rank}(D z_i) = \dim A$. We then get the following variant of Ax-Schanuel in this setting.

Formulation 5.7 *Let U' be a geodesic subvariety of $\mathbb{C}^n$. Put $X' = \exp U'$ and let A be a component of $W \cap \pi^{-1}(V)$, where $W \subset U'$ and $V \subset X'$ are algebraic subvarieties. If A is not contained in any proper geodesic subvariety of U' then*

$$\dim A \leq \dim V + \dim W - \dim X'.$$

I.e. (and as observed still more generally by Ax [4]), the components of the intersection of W and $\pi^{-1}(V)$ never have "atypically large" dimension, except when A is contained in a proper geodesic subvariety. It is convenient to give an equivalent formulation.

Definition 5.8 Fix $V \subset (\mathbb{C}^\times)^n$.

1. A *component with respect to V* is a component of $W \cap \pi^{-1}(V)$ for some $W \subset \mathbb{C}^n$.
2. If A is a component we define its *defect* by $\delta(A) = \dim \operatorname{Zcl}(A) - \dim A$.
3. A component A with respect to V is called *optimal* for V if there is no strictly larger component B w.r.t. V with $\delta(B) \leq \delta(A)$.

4. A component A w.r.t. V is called *geodesic* if it is a component of $W \cap \pi^{-1}(V)$ for some geodesic subvariety W, with $W = \mathrm{Zcl}(A)$.

Formulation 5.9   *Let $V \subset \mathbb{C}^n$. An optimal component for V is geodesic.*

We show that these two formulations are equivalent using only formal properties of weakly special subvarieties, so that the equivalence will hold in more general settings we will consider.

Proof that 5.7 implies 5.9   We assume Formulation 5.7 and suppose that the component A of $W \cap \pi^{-1}(V)$ is optimal, where $W = \mathrm{Zcl}(A)$. Suppose that U' is the smallest geodesic subvariety containing A, and let $X' = \pi(U')$. Then $W \subset U'$. Let $V' = V \cap X'$. Then A is optimal for V' in U', otherwise it would fail to be optimal for V in $\mathbb{C}^n$. Since A is not contained in any proper geodesic subvariety of U' we must have

$$\dim A \le \dim W + \dim V' - \dim X'.$$

Let B be the component of $\pi^{-1}(V')$ containing A. Then B is also not contained in any proper geodesic subvariety of U', so, by Formulation A,

$$\dim B \le \dim V' + \dim \mathrm{Zcl}(B) - \dim X'.$$

But $\dim B = \dim V'$, whence $\dim \mathrm{Zcl}(B) = \dim X'$, and so $\mathrm{Zcl}(B) = X'$, and B is a geodesic component. Now

$$\delta(A) = \dim W - \dim A \ge \dim X' - \dim V' = \delta(B)$$

whence, by optimality, $A = B$.   □

Proof that 5.9 implies 5.7   We assume Formulation 5.9. Let U' be a geodesic subvariety of $\mathbb{C}^n$, put $X' = \pi(U)$. Suppose $V \subset X', W \subset U'$ are algebraic subvarieties and A is a component of $W' \cap \pi^{-1}(V')$ not contained in any proper geodesic subvariety of U'. There is some optimal component B containing A, and B is geodesic, but since A is not contained in any proper geodesic, B must be a component of $\pi^{-1}(V')$ with $\mathrm{Zcl}(B) = U'$ and we have

$$\dim W - \dim A \ge \delta(A) \ge \delta(B) = \dim X' - \dim V$$

which rearranges to what we want.   □

6 "Ax-Lindemann"

We retain the setting $\pi : \mathbb{C}^n \to (\mathbb{C}^\times)^n$ and terminology from the previous section. A component of defect zero with respect to $V \subset X$ is then just an algebraic subvariety $W \subset \pi^{-1}(V)$. We thus have by Formulation B:

Ax-Lindemann, Form 1 *A maximal algebraic subvariety* $W \subset \exp^{-1}$ *(V) is geodesic.*

Let us explicate Form 1. If $W \subset \pi^{-1}(V)$ then we may consider W to be a component w.r.t. V. If W is not contained in any proper geodesic subvariety we find

$$\dim W \leq \dim V + \dim W - \dim X$$

so that $\dim X \leq \dim V$, i.e. $\pi(W)$ is Zariski-dense in X. Let us consider then a subvariety $W \subset U$ with $z_1, \ldots, z_n$ denoting the elements of $\mathbb{C}(W)$ induced by the coordinate functions. We get the following.

Ax-Lindemann, Form 2 *If* $z_1, \ldots, z_n \in \mathbb{C}(W)$ *are l.i.$/\mathbb{Q}$ mod $\mathbb{C}$ then the functions*

$$e^{z_1}, \ldots, e^{z_n}$$

are algebraically independent over $\mathbb{C}$.

In this form it should be clear that this is an analogue of Lindemann's theorem for algebraic functions (i.e. elements of the algebraic function field $\mathbb{C}(W)$), hence the neologism "Ax-Lindemann" to denote (retrospectively) this part of Ax-Schanuel.

However it is Form 1 that is important in the applications: for it essentially characterises the "algebraic part" of $\pi^{-1}(V)$. The algebraic part is defined in terms of real semi-algebraic subsets of $\pi^{-1}(V)$ (connected and of positive dimension). Because $\pi^{-1}(V)$ is complex analytic, it turns out that $\pi^{-1}(V)^{\mathrm{lag}}$ is in fact a union of complex algebraic varieties. By "Ax-Lindemann" it is a union of geodesic subvarieties.

We give direct proofs of the equivalence of these two forms, which are formal (and therefore will hold in more general settings).

Proof that 1 implies 2 Suppose $e^{z_1}, \ldots, e^{z_n}$ as in Form 2 are not algebraically independent over $\mathbb{C}$. So $\exp(W) \subset V$ for some proper algebraic subvariety $V \subset (\mathbb{C}^*)^n$. By Form 1, there is a geodesic W' with $W \subset W' \subset \exp^{-1}(V)$. Since V is a proper subvariety, so is W' and so there is a non-trivial equation $\sum q_i z_i = c$ that holds on W. Hence the coordinate functions $z_1, \ldots, z_n$ are linearly dependent over $\mathbb{Q}$ modulo $\mathbb{C}$. $\square$

Proof that 2 implies 1 We consider V as in the statement of Form 1, and $W \subset \exp^{-1}(V)$ maximal. Choose a maximal subset $z_i, i \in I \subset \{1, \ldots, n\}$ such that e^{z_i} are algebraically independent over $\mathbb{C}$. So all the other z_j are "geodesically dependent" on these, i.e. there is an equation $z_j = c_j + \sum_{i \in I} q_{ij}, c_j \in \mathbb{C}, q_{ij} \in \mathbb{Q}$. Since the $\exp z_i, i \in I$ are algebraically independent, we see that the geodesic subvariety T defined by the above equations for each $j \notin I$ is contained in $\pi^{-1}(V)$. By maximality $W = T$.   □

7 The modular function

Let $\mathbb{H} = \{z \in \mathbb{C} : \mathrm{Im}(z) > 0\}$ denote the complex upper half-plane. The *(elliptic) modular function* or *modular invariant* or *j-function* is a holomorphic function

$$j : \mathbb{H} \to \mathbb{C}$$

with remarkable arithmetic properties. We will describe some of its properties before briefly indicating the role this function plays in the arithmetic of elliptic curves.

In the following, various 2×2 real matrices with positive determinant act on $\mathbb{H}$ as *Mobius transformations* as follows:

$$g = \begin{pmatrix} a & b \\ c & d \end{pmatrix} \quad \text{acts by} \quad z \mapsto gz = \frac{az + b}{cz + d}.$$

The condition $\det g > 0$ ensures $g(\mathbb{H}) = \mathbb{H}$.

Firstly, j is invariant under the action by $\mathrm{SL}_2(\mathbb{Z})$. The action of $\mathrm{SL}_2(\mathbb{Z})$ on $\mathbb{H}$ has a classical fundamental domain

$$F = \{z \in \mathbb{H} : |\mathrm{Re}(z)| \leq 1/2, |z| \geq 1\}.$$

More precisely this is the closure of a true fundamental domain F^* (i.e. where each $\mathrm{SL}_2(\mathbb{Z})$ orbit is represented just once), as the transformation $z \mapsto z + 1$ identifies the vertical strips $\mathrm{Re}(z) = \pm 1/2$ and the transformation $z \mapsto -1/z$ identifies two segments of the circular boundary. A proof that F is a fundamental domain can be found in Serre [57]. Examining it yields a quantitative statement that will be important for us.

Proposition 7.1 ([48]) *Let $z \in \mathbb{H}$. The (unique) $\gamma \in \mathrm{SL}_2(\mathbb{Z})$ such that $\gamma z \in F^*$ has entries bounded by a polynomial (of degree ≤ 7) in $\max(|z|, (\mathrm{Im}(z))^{-1})$.*   □

More generally, we consider the action by $GL_2^+(\mathbb{Q})$, where the $^+$ denotes positive determinant. Scaling a matrix does not change its action, so we could reduce everything to actions by elements of $SL_2(\mathbb{R})$. However this does not preserve rationality of the entries, so it is convenient to work with $GL_2^+(\mathbb{Q})$.

For each $g \in GL_2^+(\mathbb{Q})$ we may scale the matrix until its entries are in $\mathbb{Z}$ but relatively prime. The determinant of this matrix we denote $N = N(g)$.

For each $N \geq 1$ there is a *modular polynomial*

$$\Phi_N \in \mathbb{Z}[X, Y],$$

symmetric for $N \geq 2$ ($\Phi_1 = X - Y$), such that, if $N(g) = N$,

$$\Phi_N\big(j(z), j(gz)\big) = 0,$$

i.e. the two functions $j(z), j(gz)$ are algebraically dependent (over $\mathbb{Q}$). For example, $2z = \begin{pmatrix} 2 & 0 \\ 0 & 1 \end{pmatrix} z$, and $X = j(z), Y = j(2z)$ are related by $\Phi_2(X, Y) = 0$ where

$$\Phi_2 = -X^2Y^2 + X^3 + 1488(X^2Y + XY^2) + Y^3 - 162.10^3(X^2 + Y^2)$$

$$+ 40773375XY + 8748.10^3(X + Y) - 157464.10^9.$$

More details and examples can be found e.g. in Zagier [67], Diamond-Sherman [17].

Since j is invariant under $z \mapsto z + 1$ it has a Fourier expansion, known as the *q-expansion*

$$j(z) = q^{-1} + 744 + \sum_{m=1}^{\infty} c_m q^m, \quad q = e^{2\pi i z}$$

where (it turns out) $c_i \in \mathbb{Z}$ (e.g. $c_1 = 196884$). Thus, as one goes vertically to infinity, say along $z = it$, $j(z)$ grows like $e^{2\pi t}$ and has an essential singularity at ∞, and likewise at every rational point on the real line ($SL_2(\mathbb{Z})$ acts transitively on $\mathbb{Q} \cup \{\infty\}$).

If $z \in \mathbb{H}$ with $[\mathbb{Q}(z) : \mathbb{Q}] = 2$ then $j(z)$ is algebraic, indeed it is an algebraic integer with rich arithmetical properties described by the theory of *complex multiplication of elliptic curves*. For now we mention that, if z satisfies the quadratic equation

$$az^2 + bz + c = 0$$

where $a, b, c \in \mathbb{Z}, a > 0, (a, b, c) = 1$, letting $D(z) = b^2 - 4ac < 0$ be its *discriminant* then

$$[\mathbb{Q}\big(j(z)\big) : \mathbb{Q}] = h(D)$$

where $h(D)$ is the *class number* of the corresponding quadratic order; in particular if D is square-free then the corresponding order is the ring of integers in $\mathbb{Q}(z)$, and $h(D)$ is the order of its *class group*, the (finite) group of ideal classes under composition.

By a result of Schneider (1937), there are no other $z \in \mathbb{H}$ for which z and $j(z)$ are simultaneously algebraic ("Modular Hermite-Lindemann").

The modular function satisfies a third order algebraic differential equation, but none of any smaller order (Mahler [34]). Indeed (see e.g. Bertrand-Zudilin [10])

$$j''' \in \mathbb{Q}(j, j', j''),$$

more precisely (Masser [35])

$$Sj + \frac{j^2 - 1968j + 2654208}{2j^2(j - 1728)^2}(j')^2 = 0$$

where

$$Sf = \frac{f'''}{f'} - \frac{3}{2}\left(\frac{f''}{f'}\right)^2$$

is the *Schwarzian derivative*. We have $Sf = 0$ iff $f \in SL_2(\mathbb{C})$.

Let us now say a little about how $j(z)$ arises in the theory of elliptic curves. If $\Lambda \subset \mathbb{C}$ is a lattice (discrete $\mathbb{Z}$ module of rank 2), one can create doubly periodic meromorphic functions. By scaling one can always consider such lattices to be of the form

$$\Lambda_\tau = \mathbb{Z} + \mathbb{Z}\tau, \quad \tau \in \mathbb{H}.$$

Then one forms the Weierstrass $\wp$-function $\wp_\tau(z)$ by summing a suitable simple expression over the lattice being careful that it converges. It is doubly periodic and has double poles at the lattice points.

Its derivative is also Λ_τ periodic, and by taking suitable combinations one can eliminate the pole. The resulting function must vanish and one finds a relation of the form

$$\wp'^2 = 4\wp^2 - g_2(\tau)\wp(z) - g_3(\tau)$$

for suitable $g_2(\tau), g_3(\tau) \in \mathbb{C}$. Thus the map

$$z \mapsto (\wp, \wp')$$

maps $\mathbb{C}/\Lambda_\tau$ to a complex algebraic curve (including the point at ∞, giving a smooth projective curve). This is an *elliptic curve*, which we denote E_τ, and inherits a group law from the additive structure on $\mathbb{C}/\Lambda_\tau$. On E_τ the group law is given by some rational functions.

An elliptic curve is determined up to isomorphism over $\mathbb{C}$ by its j-invariant

$$j(\tau) = 1728 \frac{g_2^3(\tau)}{g_2^3(\tau) - 27g_3^2(\tau)}.$$

The isomorphism can also be read in the lattices: τ_1, τ_2 give isomorphic curves if they are equivalent under $SL_2(\mathbb{Z})$ which amounts to change of basis and rescaling. Thus the $SL_2(\mathbb{Z})$ invariance of j.

Actions by $g \in GL_2^+(\mathbb{Q})$ correspond to *isogenies* (homomorphisms with finite kernel) between elliptic curves. This can be seen as taking the quotient of a given curve by some (cyclic) subgroup, and so the j-invariant of the quotient has some algebraic relation to the given curve.

It is an elaborate and beautiful theory. See Diamond and Shurman [17], Zagier [67].

8 Modular Schanuel Conjecture

Definition 8.1 A point $z \in \mathbb{H}$ is called *special* if $[\mathbb{Q}(z) : \mathbb{Q}] = 2$.

Our principle now is that "numbers defined using the j-function should be as algebraically independent as permitted by the modular relations and the special values". We introduce a suitable "independence" notion.

Definition 8.2 Elements $z_1, \ldots, z_n \in \mathbb{H}$ are called $GL_2^+(\mathbb{Q})$-*independent* if the z_i are not special and there are no relations

$$z_i = g z_j, \quad i \neq j, \quad g \in GL_2^+(\mathbb{Q}).$$

In fact the special points are fixed points, so one could rephrase this as "no non-trivial $GL_2^+(\mathbb{Q})$ relations between the z_i", the trivial ones being $z_i = 1 z_i$. Note also that the relations are pairwise: if a set of n elements is dependent then one or two of them are already dependent. This is the hallmark of a "trivial pregeometry".

A first formulation might be the following.

Conjecture 8.3 Suppose $z_1, \ldots, z_n$ are $GL_2^+(\mathbb{Q})$- independent. Then

$$\text{tr.d.}\big(z_1, \ldots, z_n, j(z_1), \ldots, j(z_n)\big) \geq n.$$

Schneider's result gives this for $n = 1$; for $n \geq 2$ it is not known. Even the "Lindemann" statement (i.e. with z_i algebraic) is open. Many things are known beyond Schneider's result, for which I refer to Diaz [18] and Nesterenko [43].

The above conjecture does not take into account the derivatives of j. These fit into a much bigger conjectural picture which includes elliptic functions as well as modular ones (and the higher dimensional analogues), namely the *generalised period conjecture* of Grothendieck-André: see André [2], Bertolin [7]. The following may be deduced from an explication of this conjecture in the case of one-dimensional "motives" in [7].

Modular Schanuel Conjecture 8.4   Suppose $z_1,\ldots,z_n$ are $\mathrm{GL}_2^+(\mathbb{Q})$-independent. Then

$$\mathrm{tr.d.}\bigl(z_1,\ldots,z_n,j(z_1),\ldots,j(z_n),j'(z_1),\ldots,j'(z_n),j''(z_1),\ldots,j''(z_n)\bigr) \geq 3n.$$

This does not reflect some transcendence properties of the derivatives at special points, but it is sufficient for our purposes here.

9 "Modular Ax-Schanuel"

We work in the complex setting rather than digressing on formulating a "Modular Ax-Schanuel" in a differential field. We consider

$$\pi : \mathbb{H}^n \to \mathbb{C}^n, \quad \pi(z_1,\ldots,z_n) = (j(z_1),\ldots,j(z_n)).$$

Let $A \subset U$ be a complex analytic subvariety of some open $U \subset \mathbb{H}^n$, with the coordinate functions $z_1,\ldots,z_n$ and $j(z_1),\ldots,j(z_n)$ meromorphic on A, and with derivations $\{D_k\}$ induced by differentiation w.r.t. z_k such that $\mathrm{rank}(D_k z_\ell) = \dim A$, the rank being over the field of meromorphic functions on A.

Definition 9.1   The functions $z_1,\ldots,z_n$ on A are called *geodesically independent* if no z_i is constant and there are no relations $z_k = gz_\ell$ where $k \neq \ell$ and $g \in \mathrm{GL}_2^+(\mathbb{Q})$.

The following conjecture might be considered the analogue of "Ax-Schanuel" for the *j*-function.

"Modular Ax-Schanuel" Conjecture 9.2   In the above setting, suppose that the z_i are geodesically independent. Then

$$\mathrm{tr.d.}_{\mathbb{C}}\mathbb{C}(z_1,\ldots,z_n,j(z_1),\ldots,j(z_n)) \geq n + \dim A.$$

This conjecture is open beyond some special cases described below (including further below in §16); it is of intrinsic interest, but also very useful in addressing Zilber-Pink problems (see §15).

We pursue now geometric formulations analogous to those obtained earlier for the exponential function, and they will take exactly the same form. To frame these we need a definition of "geodesic subvariety", but we also need to pause on the meaning of an "algebraic subvariety" of $\mathbb{H}^n$. We can map $\mathbb{H}^n$

to the product Δ^n of open unit discs by an invertible algebraic map, whence one sees that there can be no positive dimensional algebraic varieties contained inside $\mathbb{H}^n$.

Definition 9.3 By a "subvariety" of $\mathbb{H}^n$ we mean an irreducible (in the complex analytic sense) subvariety of $W \cap \mathbb{H}^n$ for some algebraic subvariety $W \subset \mathbb{C}^n$.

Definition 9.4 A subvariety $W \subset \mathbb{H}^n$ is called *geodesic* if it is defined by some number of equations of the forms

$$z_i = c_i, \quad c_i \in \mathbb{C}; \quad z_k = g_{k\ell} z_\ell, \quad g \in \mathrm{GL}_2^+(\mathbb{Q}).$$

These are the "weakly special subvarieties" in the Shimura sense. The word "geodesic" is adopted from Moonen [41] who shows that, in a Shimura variety, the weakly special subvarieties are the "totally geodesic" ones. I wanted a word that gave a readable "suchly independent" phrase in analogy with "linearly independent" and "algebraically independent".

Since we have defined the "weakly special subvarieties" it is opportune to define the special ones.

Definition 9.5

1. A *special point* in $\mathbb{H}^n$ is a tuple of special (i.e. quadratic) points.
2. A *special subvariety* in $\mathbb{H}^n$ is a weakly special subvariety containing a special point; equivalently, the fixed coordinates c_i above are all special.
3. The images under π of these are the *special subvarieties* in $\mathbb{C}^n$.

We now define components, their defects, and optimal components exactly as before and find that the conjecture above implies the following two formulations of a "Weak Modular Ax-Schanuel" conjecture, which are equivalent by exactly the same proofs given previously.

Formulation 9.6 *Let U' be a geodesic subvariety of $\mathbb{H}^n$. Put $X' = \exp U'$ and let A be a component of $W \cap \pi^{-1}(V)$, where $W \subset U'$ and $V \subset X'$ are algebraic subvarieties. If A is not contained in any proper geodesic subvariety of U' then*

$$\dim A \leq \dim V + \dim W - \dim X'.$$

Formulation 9.7 *Let $V \subset \mathbb{C}^n$. An optimal component for V is geodesic.*

Formulation 9.7 is the form that is needed to tackle Zilber-Pink problems using o-minimality and point-counting. However, a true "Modular Ax-Schanuel" should take into account the derivatives of j.

Conjecture 9.8 (Modular Ax-Schanuel with derivatives)   In the setting of "Modular Ax-Schanuel" above, if z_ℓ are geodesically independent then

$$\text{tr.d.}_{\mathbb{C}}\mathbb{C}(z_1,\ldots,z_n,j(z_1),\ldots,j(z_n),j'(z_1),\ldots,j'(z_n),j''(z_1),\ldots,j''(z_n))$$
$$\geq 3n + \dim A.$$

The "geodesic independence" condition is evidently equivalent in the complex setting (in analogy with Ax [3]) to: the $j(z_i)$ are "modular independent", i.e. non-constant and no relation $\Phi_N(j(z_k),j(z_\ell)) = 0$ holds for $k \neq \ell, N \geq 1$.

•

10 "Modular Ax-Lindemann"

We retain the setting of the previous section, and consider $V \subset \mathbb{C}^n$.

"Modular Ax-Lindemann" Form 1   *A maximal algebraic subvariety $W \subset j^{-1}(V)$ is geodesic.*

"Modular Ax-Lindemann" Form 2   Let $W \subset \mathbb{C}^n$ with $W \cap \mathbb{H}^n \neq \emptyset$. Suppose that the coordinate functions $z_1,\ldots,z_n \in \mathbb{C}(W)$ are geodesically independent. Then the functions

$$j(z_1),\ldots,j(z_n),$$

defined locally on W, are algebraically independent over $\mathbb{C}$.

These two formulations are equivalent, by variants of the proofs for the exponential case (**Exercise**). It is Form 1 that arises in the o-minimal approach to the André-Oort conjecture for products of modular curves. It is proved in [48], and we sketch the proof later. A version "with derivatives" is established in [49].

11 The general setting

Both settings described above: $\exp : \mathbb{C}^n \to (\mathbb{C}^*)^n$, $j : \mathbb{H}^n \to \mathbb{C}^n$ fit into a bigger picture

$$\pi : U \to X$$

where X is a *Shimura* or *mixed Shimura variety* (see Pink [55, 56], or Daw [16] in this volume), and U is (essentially) its universal cover.

The prototypical Shimura varieties are modular varieties. For example $\mathbb{C}$, as the j-line, is the moduli space of elliptic curves up to isomorphism over $\mathbb{C}$. The higher dimensional analogues are the *Siegel moduli spaces* $\mathcal{A}_g$ which parameterise (principally polarised) abelian varieties of dimension g, i.e. g-dimension complex tori which admit an algebraic structure (when $g \geq 2$ not all do). The dimension of $\mathcal{A}_g$ is $g(g+1)/2$. One has

$$\pi_g : \mathbb{H}_g \to \mathcal{A}_g$$

where $\mathbb{H}_g$ is the *Siegel upper half space*, and the uniformisation (which is given by *Siegel modular forms*) is invariant under $\mathrm{Sp}_{2g}(\mathbb{Z})$. See e.g. van der Geer [19].

Each point $x \in \mathcal{A}_g$ parameterises an abelian variety A_x; the corresponding *mixed* Shimura variety consists of $\mathcal{A}_g$ fibered by the A_x. The simplest example is given by the Legendre family of elliptic curves, that is the elliptic surface

$$y^2 = x(x-1)(x-\lambda)$$

considered as a family of elliptic curves, one for each $\lambda \in \mathbb{C} - \{0, 1\}$ fibered over the λ-line.

Maybe here is the point to mention that $\mathbb{H}$ is not the universal cover of $\mathbb{C}$. The covering by j is ramified at two points whose pre-images are fixed by elements of $\mathrm{SL}_2(\mathbb{Z})$, namely $j(i)$ and $j(\rho)$. But if one takes a suitable finite index (congruence) subgroup one has no such points: the covering

$$\lambda : \mathbb{H} \to \mathbb{C} - \{0, 1\}$$

associated with the Legendre family is universal, and the corresponding congruence subgroup is isomorphic to the free group on two generators.

In general, mixed Shimura varieties arise as quotients of symmetric hermitian domains by suitable arithmetic subgroups of their group of biholomorphic self-maps. They have the structure of an algebraic variety. Each mixed Shimura variety X has a collection $\mathcal{S} = \mathcal{S}_X$ of "special subvarieties" and a larger collection $\mathcal{W} = \mathcal{W}_X$ of "weakly special subvarieties", which is what I have termed "geodesic".

Shimura varieties are the setting for an arithmetic conjecture called the "Andre-Oort conjecture". This fits into the much broader "Zilber-Pink" conjecture in the setting of mixed Shimura varieties, which concerns the interaction between subvarieties $V \subset X$ and the collection of "special subvarieties" (see §15).

In the approach to these conjectures via o-minimality, suitable functional transcendence statements are a key ingredient. In particular, to carry out this approach to prove AO for a Shimura variety X one requires:

"Ax-Lindemann" Conjecture for X 11.1 *Let $V \subset X$. A maximal algebraic subvariety $W \subset \pi^{-1}(V)$ is weakly special.*

Klingler-Ullmo-Yafaev have recently announced [30] a proof of this for all Shimura varieties: Tsimerman and I [52] proved it for $\mathcal{A}_g$, building in part on Ullmo-Yafaev's proof [64] for all *compact* Shimura varieties (when there are no cusps in the fundamental domain, the quotient is a compact, i.e. projective, variety). In proving this theorem (using o-minimality) they also established the definability of the uniformisation restricted to a fundamental domain, extending the work of Peterzil-Starchenko [47] who did it for $\mathcal{A}_g$ (indeed for the mixed Shimura variety associated with $\mathcal{A}_g$). The extension to mixed Shimuras may not be far away.

This means, by work of Ullmo [60], that a full proof of AO is now reduced to (1) a statement about Galois orbits of special points being "large", and (2) a statement that the height of a pre-image of a special point is "not too large". For $\mathcal{A}_g$, the latter was proved in [51]; for $g \leq 6$, the Galois lower bound is known due to Tsimerman ([58]; under GRH it is known for all g by Tsimerman and independently by Ullmo-Yafaev [62]). More generally one expects the following.

Weak Ax-Schanuel Conjecture 11.2 Let X be a (mixed) Shimura variety and $V \subset X$. An optimal component for $V \subset X$ is weakly special.

For a still more general setting see Zilber [73].

12 Exponential Ax-Lindemann via o-minimality

We give a proof of "Ax-Lindemann" using o-minimality and point-counting, to motivate the proof of the modular analogue which follows.

For a proof of the full Ax-Schanuel statement via o-minimality and point-counting see Tsimerman [59], in this volume. We consider

$$\exp : \mathbb{C}^n \to (\mathbb{C}^*)^n, \quad V \subset (\mathbb{C}^*)^n.$$

The complex exponential is definable when restricted to a fundamental domain for the $2\pi i \mathbb{Z}$ action (by translation) on $\mathbb{C}$. We take say

$$F = \{z \in \mathbb{C} : 0 \leq \mathrm{Im}(z) < 2\pi i\}.$$

Then $\exp$ is definable on F^n, and we let

$$Z = \exp^{-1}(V) \cap F^n,$$

which is also definable.

Theorem 12.1   *A maximal algebraic subvariety $W \subset \exp^{-1}(V)$ is geodesic.*

Idea 12.2

1. The action of $(2\pi i\mathbb{Z})^n$ on F^n divides $\mathbb{C}^n$ into fundamental domains γF^n, where $\gamma \in (2\pi i\mathbb{Z})^n$. We find that W is "present" in "many" of them. Then the suitable translation of these pieces back to F^n belongs to $\exp^{-1}(V)$.
2. The $\gamma \in (2\pi i\mathbb{Z})^n$ for which W is "present" in γF belong to a certain definable subset of $(2\pi i\mathbb{R})^n$ for which the corresponding translate of W is contained in $\exp^{-1}(V)$, and which thus contains "many" rational points. By the Counting Theorem, this set contains positive dimensional semi-algebraic families of translates of W.
3. Consider such a family of translations, say with a real parameter t. If the union over this family of translations is bigger than W, we could "complexify" the parameter and get a complex variety W' containing W but of bigger dimension. This contradicts our assumption that W is maximal. So these translations must translate W along itself. This forces W to be linear and even to be a coset of a rational subspace.

Proof   We suppose $W \subset \exp^{-1}(V)$ is maximal, say of dimension k. We can suppose that $z_1, \ldots, z_k$ are independent functions on W, and that the other variables depend algebraically on them

$$z_\ell = \psi_\ell(z_1, \ldots, z_k), \quad \ell = k+1, \ldots, n.$$

Of course these algebraic functions will have some branching, but locally at smooth points they are functions and can be analytically continued throughout $z_1, \ldots, z_k$-space avoiding some lower-dimensional branching locus.

We will write below z for the tuple of "free" variables $(z_1, \ldots, z_k)$, and ψ for the tuple of functions $(\psi_{k+1}, \ldots, \psi_n)$.

Fix some small product of discs $U \subset \mathbb{C}^k$ in the $z_1, \ldots, z_k$-variables such that the ψ_ℓ are all unbranched at points

$$(z + 2\pi it) = (z_1 + 2\pi it_1, z_2 \ldots, z_k)$$

for $(z_1, \ldots, z_k) \in U$ and all sufficiently large real t_1 (this is true generically). By the periodicity of $\exp$, any translation of W by integer multiples of $2\pi i$ on the coordinates is again inside $\exp^{-1}(V)$.

But we are going to use definability, so can only make use of $\exp$ on finitely many fundamental domains. We will just use F^n.

For any integer t, there exists a unique integer vector

$$m'(t) = (m_{k+1}, \ldots, m_n)$$

such that the graph on U of

$$\psi(z + 2\pi i t) - 2\pi i m'(t)$$

intersects Z in a set of real dimension $2k$ (which is its full real dimension).

For any $m' \in \times \mathbb{R}^{n-k}$ and $t \in \mathbb{R}$ we let

$$W(U, m', t)$$

denote the graph on U of the functions

$$\psi(z + 2\pi i t) - 2\pi i m'.$$

Fixing U we consider now the definable set

$$Y = \{(m', t) \in \mathbb{R}^{n-k} \times \mathbb{R} : \dim_{\mathbb{R}} \left(W(U, m', t) \cap Z \right) = 2k\}.$$

Since the functions ψ_ℓ have polynomial growth in t, the components of $m'(t)$ are bounded by some polynomial in t. Therefore Y contains "many" rational (in fact integer) points.

Therefore Y contains semi-algebraic curves which contain arbitrarily large finite numbers of integer points, which seems to give us a positive family of translates of W contained in $\exp^{-1}(V)$ if we "complexify" the parameter t locally.

But W is maximal, so we must be just translating W along itself, in particular we have for suitable integers $s_1 \neq t_1$ and integer vectors $m'(s_1), m'(t_1)$

$$(*) \qquad \psi_\ell(z_1 + 2\pi i s_1, z_2, \dots, z_k) - \psi_\ell(z_1 + 2\pi i t_1, z_2, \dots, z_k)$$

$$= 2\pi i m'(s_1) - 2\pi i m'(t_1)$$

holding for all ℓ *identically in z*.

Fix $(z_2, \dots, z_k)$. Differentiating with respect to z_1, we see that

$$\psi'_\ell(z_1 + 2\pi i s_1, z_2, \dots, z_k) - \psi'_\ell(z_1 + 2\pi i t_1, z_2, \dots, z_k) = 0.$$

The algebraic function $\psi'_\ell(z_1, z_2, \dots, z_k)$ (as a function of z_1, the other z_i being fixed) with a period must be constant. So we have

$$\psi_\ell(z_1, z_2, \dots, z_k) = q(z_2, \dots, z_k)z_1 + r(z_2, \dots, z_k).$$

Since we have integer points, if we go back to $(*)$ the coefficient $q(z_2, \dots, z_k)$ must be rational. But then $q(z_2, \dots, z_k)$, which is an algebraic function, must be constant (could use definability here) and we have

$$\psi_\ell(z_1, z_2, \dots, z_k) = q_{\ell 1} z_1 + r(z_2, \dots, z_k), \qquad q_{\ell 1} \in \mathbb{Q}.$$

We repeat the argument for all the free variables to find that

$$\psi_\ell = r_\ell + \sum q_{\ell i} z_i, \quad q_{\ell i} \in \mathbb{Q}, r_\ell \in \mathbb{C}, \quad \ell = k+1, \ldots, n$$

and so W is geodesic. $\qquad\qquad\qquad\qquad\qquad\qquad\qquad\qquad\qquad\square$

13 Modular Ax-Lindemann via o-minimality

We sketch a proof of Modular Ax-Lindemann via o-minimality and point-counting, essentially following the argument in [48]. This follows the previous argument, with just a few minor additional technicalities due to the boundary of $\mathbb{H}^n$ and the more intricate group action. We now consider

$$j : \mathbb{H}^n \to \mathbb{C}^n, \quad V \subset \mathbb{C}^n.$$

Definability 13.1 Let F be the standard fundamental domain for the $\mathrm{SL}_2(\mathbb{Z})$ action on $\mathbb{H}$, as described above. Then $j|_F : F \to \mathbb{C}$ is definable in $\mathbb{R}_{\mathrm{an\,exp}}$. (Follows from the q-expansion; Peterzil-Starchenko [46] observed this while proving definability for the Weierstrass $\wp$ function as a function of both variables.)

Theorem 13.2 *A maximal algebraic subvariety $W \subset j^{-1}(V)$ is geodesic.*

Idea 13.3 The same.

Proof We suppose $W \subset j^{-1}(V)$ is maximal, with $\dim W = k$, and that, locally on some region $D \subset \mathbb{C}^k$, we may take $z_1, \ldots, z_k$ as independent variables and parameterise W by

$$z_\ell = \phi_\ell(z_1, \ldots, z_k), \quad \ell = k+1, \ldots, n.$$

If $g \in \mathrm{SL}_2(\mathbb{Z})^n$ then gW is also a maximal algebraic subvariety. It is locally parameterised by

$$z_\ell = g_\ell \phi_\ell(g_1^{-1} z_1, \ldots, g_k^{-1} z_k), \quad \ell = k+1, \ldots, n$$

on $(g_1, \ldots, g_k)D$.

We can then analytically continue these functions, perhaps with some branching, remaining inside $\mathbb{H}^n$ (and hence within $j^{-1}(V)$), until some free or dependent variable runs into its real line.

For example, keeping $z_2, \ldots, z_k$ in a small neighbourhood, we can analytically continue the functions in z_1 up to the real boundary unless some ϕ_ℓ becomes real. This ϕ_ℓ then depends on z_1 over $z_2, \ldots, z_k$, and we can exchange

z_1 and z_ℓ, and now we have a parameterisation that goes up to the boundary of z_1.

This gives a subregion of $\mathbb{H}^n$ bounded by loci where some $z_1, \ldots, z_k$ or some $\phi_\ell(z_1, \ldots, z_k)$ becomes real, in particular including some "disc" U_1 where z_1 becomes real, the other free variables remain away from their real lines (so it is a product of a half-disc in z_1, and discs in the other free z_i). Some of the dependent z_ℓ may also be real on U_1, others not: we can move to a smaller "disc" so that such dependent variables are either real on all of U_1, or are away from their real lines – and contained in a single fundamental domain for $\mathrm{SL}_2(\mathbb{Z})$.

Again moving to a smaller "disc" if necessary we can assume that all the ϕ_ℓ are regular and non-branching.

We let Φ denote the tuple $(\phi_{k+1}, \ldots, \phi_n)$ and put

$$W_1 = \{(u, \Phi(u)) : u \in U_1\} \subset W,$$

a definable (even semi-algebraic) set.

The point will be that in the z_1 half disc there are infinitely many fundamental domains but the variables away from their real lines will be confined to finitely many fundamental domains.

Fix a fundamental domain F_1 inside the z_1 half disc, and a rational point a/c with $(a, c) = 1$ on the boundary of this half disc. Take a matrix

$$g_0 = \begin{pmatrix} a & b \\ c & d \end{pmatrix} \in \mathrm{SL}_2(\mathbb{Z})$$

and write

$$g_0(t) = \begin{pmatrix} a & b+t \\ c & d+t \end{pmatrix}, \quad t \in \mathbb{R}.$$

For large real t and $z \in F_1$, $g_0(t)z$ is in the half disc. Let

$$G_0 = \{g \in \mathrm{SL}_2(\mathbb{R})^n : g_1 = g_0(t), \text{ some } t, g_i = 1, i \le 2 \le k\}$$

with no restriction on $g_{k+1}, \ldots, g_\ell$. This set is clearly definable.

For any definable $G' \subset \mathrm{SL}_2(\mathbb{R})^n$, $W' \subset W$ (of full complex dimension k say) and $Z' \subset j^{-1}(V)$ the set

$$R(G', W', Z') = \{g \in G' : \dim_\mathbb{R}(gW' \cap Z) = 2k\}$$

is definable. Further, for any such g we have $gW \subset j^{-1}(V)$ by dimensional considerations and analytic continuation.

Consider the definable set

$$R(G_0, Y_1, Z).$$

For large t the action by $g_0(t)$ keeps part of the z_1 half disc within itself. For any such t we may find elements of $SL_2(\mathbb{Z})$ to bring the relevant coordinates to F, for t a large integer this will give an element of $R(G_0, Y_1, Z)$, and the size of the group element is bounded by some polynomial in t (by Proposition in §7). So $R(G_0, Y_1, Z)$ has "many" integer points, and by the Counting Theorem there are semi-algebraic subsets with arbitrarily large finite numbers of integer points.

Now maybe all these sets have fixed t. Then we can find an integer t and a positive dimensional set of translations, and hence a smooth one-dimensional set of translations contained in $j^{-1}(V)$ containing an integer point. But the integer translation of W is maximal, and is contained in a larger family by complexifying the parameter.

So we have semi-algebraic sets with many integer points and variable t. By the maximality, these translates parameterise the same translate of W.

Now we observe that the dependent variables away from their boundaries did not need to move. Therefore these variables do not depend on z_1.

For the other variables, we get identities (using two integer points on the same algebraic set where the translate is constant) of the form

$$\phi(gz) = h\phi(z), \quad \phi(z) = \phi_\ell(z, z_2, \ldots, z_k), \quad g, h \in SL_2(\mathbb{Z}).$$

We know that g is of the form

$$g_0(s)g_0(t)^{-1} = \begin{pmatrix} 1 - ac(s - t) & a^2(s - t) \\ -c^2(s - t) & 1 + ac(s - t) \end{pmatrix}$$

and so is parabolic with fixed point a/c, and we get such identities for every $a/c \in I$, the real boundary of the z_1 half disc.

Now there is an "end-game" to show that $\phi \in GL_2^+(\mathbb{Q})$. This part gets more conceptual in the various generalisations [64, 52, 30]: using monodromy considerations one shows essentially that W is an orbit of the group that stabilises it. Here I use elementary arguments.

1. $\phi \in SL_2(\mathbb{C})$

The following argument is different to the argument in [48] and to the alternative argument offered in [49].

We have $P(x, \phi(x)) = 0$ for some irreducible $P \in \mathbb{C}[X, Y]$. We have infinitely many parabolic g with distinct (real) fixed points for which we have an identity

$$\phi(gz) = h\phi(z).$$

This identity continues to hold wherever we may continue ϕ. If x_g is the fixed point of g then $y_g := \phi(x_g)$ is fixed by h, and there are infinitely many distinct y_g, even with ϕ pre-images distinct from branch points of ϕ.

If $\phi(x) = y_g$, then also $\phi(gx) = y_g$. Then x is pre-periodic under g, but since g is parabolic it has no pre-periodic points other than its unique fixed point. So for such y_g there is only one x_g (the fixed point) with $\phi(x_g) = y_g$.

Since this holds for infinitely many distinct y_g, P must be linear in X. Exchanging roles (Steinitz exchange), it is also linear in Y. So ϕ is a fractional linear transformation.

2. $\phi \in SL_2(\mathbb{R})$

Because it preserves the real line (**Exercise**).

3. ϕ is independent of $z_2, \ldots, z_k$

Because there is no non-constant holomorphic function (here complex algebraic) to $SL_2(\mathbb{R})$. E.g. the image of 0 must be a real function of a complex variable.

So the dependencies are all by fixed elements of $SL_2(\mathbb{R})$, and extend throughout all $\mathbb{H}^k$. And so we get identities for all rational a/c (and so even all $r \in \mathbb{R}$).

4. $\phi \in GL_2^+(\mathbb{Q})$

Some elementary work. Write $\phi = \begin{pmatrix} A & B \\ C & D \end{pmatrix}$ where $AD - BC = 1$. We must show that the ratios of entries are all rational (so it is in the image of $GL_2^+(\mathbb{Q})$).

We can take g with $a = 1, c = 0$. Write $u = (s - t)$. Then for some $\lambda \in \mathbb{R}$, $h \in GL_2^+(\mathbb{Q})$ we have

$$\phi g \phi^{-1} = \begin{pmatrix} 1 - uAC & uA^2 \\ -uC^2 & 1 + uAC \end{pmatrix} = \lambda h$$

for suitable (many) integer choices of u. If $C = 0$ we see that $A^2 \in \mathbb{Q}$ and then $AD = 1$ implies $A/D \in \mathbb{Q}$. Similarly, $A = 0$ implies $B/C \in \mathbb{Q}$. Otherwise $(A, C \neq 0)$ we have $A^2/C^2 \in \mathbb{Q}$ and $(1 - uAC)/C^2 \in \mathbb{Q}$, for many different u, giving $A/C \in \mathbb{Q}$. Taking $a = 0, c = 1$ we get similarly

$$\begin{pmatrix} 1 - uBD & uB^2 \\ -uD^2 & 1 + uBD \end{pmatrix} = \lambda h.$$

Now $B = 0$ leads to $A/D \in \mathbb{Q}$, $D = 0$ leads to $B/C \in \mathbb{Q}$ and otherwise $(B, D \neq 0)$ we have $B/D \in \mathbb{Q}$.

Now suppose $C = 0$, so we have $A/D \in \mathbb{Q}$. If $B = 0$ we have the required form. We cannot have $D = 0$, so $B \neq 0$ gives $B/D \in \mathbb{Q}$ and we have again the right form. Similarly, if any of $A, B, C, D = 0$ we get the right form: If $B = 0$ we have $A/D \in \mathbb{Q}$. If $C = 0$ we are done. We can't have $A = 0$, so if $C \neq 0$ we get $A/C \in \mathbb{Q}$ and are done.

So we may assume all $A, B, C, D \neq 0$. We have $A/C = q, B/D = r \in \mathbb{Q}$ and ϕ is up to scaling

$$\begin{pmatrix} 1 & \alpha \\ q & r\alpha \end{pmatrix}, \quad r \neq q, \alpha \in \mathbb{R}.$$

Then

$$\psi = \begin{pmatrix} 1 & 0 \\ -q & 1 \end{pmatrix} \phi = \begin{pmatrix} 1 & \alpha \\ 0 & (r-q)\alpha \end{pmatrix}$$

satisfies identities of the same kind (with the same g), but now there is a zero entry, so we are done! $\qquad\qquad\square$

14 SC and CIT

Boris Zilber's work on the model theory of the exponential function led him to formulate ([70, 71]) an arithmetic conjecture which he called CIT: "Conjecture on Intersections with Tori".

In the language of exponential fields one cannot formulate SC in a first order way. One can list all subvarieties $V \subset \mathbb{C}^{2n}$ defined over $\mathbb{Q}$ and of dimension $\dim V < n$. Then

$$\text{tr.d.}(z, e^z) < n, \quad z = (z_1, \ldots, z_n), \quad e^z = (e^{z_1}, \ldots, e^{z_n})$$

just means that (z, e^z) lies on one of these V and one could aspire to go through them asserting: *"If $(z, e^z) \in V$ then..."*. However one cannot assert that the coordinates of z are l.i over $\mathbb{Q}$ in a first order way, as this requires a quantification over $\mathbb{Q}$. One could do this, however, if for each such V only finitely many such linear dependencies arise: one could then just write them out explicitly.

But one must be a bit careful: the assertion *"Let $V \subset \mathbb{C}^{2n}$ be defined over $\mathbb{Q}$. There exists finitely many non-trivial linear forms $L(z_1, \ldots, z_n)$ with integer coefficients such that if*

$$(z, e^z) = (z_1, \ldots, z_n, e^{z_1}, \ldots, e^{z_n}) \in V$$

then

$$L(z_1, \ldots, z_n) = 0,$$

for (at least) one of these forms" is simply false.

Example 14.1 *Take $V \subset \mathbb{C}^3 \times \mathbb{C}^3$ defined by*

$$z_1 z_2 = z_3^2, \quad w_1 = 1, w_2 = 1, w_3 = 1.$$

So dim $V = 2$. *If* $k_1 k_2 = k_3^2$ *and* $z_\ell = 2\pi i k_\ell$ *then*

$$(z_1, z_2, z_3, e^{z_1}, e^{z_2}, e^{z_3}) \in V,$$

but these points are not contained in finitely many rational subspaces (they all lie in some proper rational subspace though!).

The right statement is a variant of this:

Uniform Schanuel Conjecture 14.2 (USC; [71]) Let $V \subset \mathbb{C}^{2n}$ be a closed algebraic set defined over $\mathbb{Q}$ with dim $V < n$. There exists a finite set $\mu(V)$ of proper $\mathbb{Q}$-linear subspaces of $\mathbb{C}^n$ such that if

$$(z_1, \ldots, z_n, e^{z_1}, \ldots, e^{z_n}) \in V$$

then there is $M \in \mu(V)$ and $\bar{k} \in \mathbb{Z}^n$ and such that $(z_1 + 2\pi i k_1, \ldots, z_n + 2\pi i k_n) \in M$. Moreover if M is codimension 1 (in $\mathbb{C}^n$) then $k = 0$.

For more on the model theory of exponentiation see [69, 28]. For the present purposes I want to work with a weaker version of USC:

Weak SC 14.3 If tr.d.$(z, e^z) < n$ then the coordinates of e^z are multiplicatively dependent.

Uniform Weak SC 14.4 Let $V \subset \mathbb{C}^{2n}$ defined over $\mathbb{Q}$ with dim $V < n$. There is a finite set $K = K(V)$ of non-trivial integer tuples $k \in \mathbb{Z}^n$ such that if $(z, e^z) \in V$ then $\prod \exp(z_i k_i) = 1$ for some $k \in K$.

Now we formulate "CIT". We consider algebraic subgroups of $X = (\mathbb{C}^*)^n$. These are subvarieties defined by some number of equations of the form

$$\prod_{i=1}^{n} x_i^{k_i} = 1$$

for vectors $k = (k_1, \ldots, k_n)$ of integers. These can be reducible (e.g. $x_1^2 = 1$ in $\mathbb{C}^*$), and they decompose into finitely many irreducible subvarieties which are called *tori* if they are subgroups, or cosets of tori by torsion points, called *torus cosets*, generally.

We will also call torus cosets *special subvarieties*, and denote the collection of them by $\mathcal{S} = \mathcal{S}(X)$. This is a countable collection.

Now two algebraic subvarieties $V, W \subset X$ generically intersect in an algebraic set whose components have dimension

$$\dim V + \dim W - \dim X$$

by simple "counting conditions" (i.e. codim V conditions are required to be on V, codim W conditions to be on W). It is a basic fact that such components can never have smaller dimension than this (see e.g. Mumford [42, 3.28]); but it can be bigger.

Definition 14.5 Let $V \subset X = (\mathbb{C}^*)^n$.

1. A component $A \subset V \cap T$, where $T \in \mathcal{S}$, is called *atypical* if

$$\dim A > \dim V + \dim T - \dim X.$$

2. Denote by

$$\mathrm{Atyp}(V) = \bigcup A$$

the union of all atypical components of $V \cap T$ over all $T \in \mathcal{S}$.

Thus $\mathrm{Atyp}(V)$ is potentially a countable union.

Conjecture 14.6 (CIT) For $V \subset (\mathbb{C}^*)^n$, $\mathrm{Atyp}(V)$ is a finite union.

Otherwise put: V contains only finitely many *maximal* atypical components.

Remarks 14.7

1. Zilber [70, 71] stated the conjecture for semi-abelian ambient varieties, and for V defined over $\mathbb{Q}$, which is what is needed for the SC application.
2. Zilber showed that CIT for (semi-)abelian varieties implies the "Mordell-Lang conjecture" (a theorem of Faltings, Raynaud, Vojta, Faltings, McQuillan), including the Manin-Mumford conjecture (Raynaud), and exponential CIT implies the multiplicative versions (Mann-Lang-Liardet-Laurent).
3. The same conjecture in the exponential setting (in a different formulation) was stated by Bombieri-Masser-Zannier ([13], for $V/\mathbb{C}$). They earlier proved partial results for curves [12]. They later proved [15] that the various formulations were equivalent in the exponential case, and that CIT/$\mathbb{Q}$ implies CIT/$\mathbb{C}$.
4. Exponential CIT is open. Various partial results are known: including a complete result for curves; see Bombieri-Masser-Zannier [12, 13, 14], Maurin [39, 40], Habegger [20, 21]; and [11].
5. The same kind of conjecture was formulated (again independently) by Pink [55, 56] in the setting of "mixed Shimura varieties". Apparently his object was to find a unifying statement including the Mordell-Lang on the "semi-abelian side" and André-Oort [1, 44, 29, 61] on the Shimura side. See Zannier's book [68].

Zilber [70, 71] proves the following theorem.

Theorem 14.8 *SC + CIT implies USC.* □

I will prove this for the weak uniform version adapting the proof in [71].

Theorem 14.9   *SC + CIT implies UWSC.*

Proof   We consider some $V \subset \mathbb{C}^{2n}$ defined over $\mathbb{Q}$ and of dimension dim $V <$ n. We let W be the projection of V onto the second $\mathbb{C}^n$ factor, d the dimension of the generic fibre of this projection, and $V' \subset V$ the proper subvariety where the fibre dimension exceeds d.

According to CIT, there is a finite set of torus cosets $S_1, \ldots, S_\ell$ whose atypical components with W contain all atypical components.

Now suppose $(z, e^z) \in V$. According to SC, z lies in some proper rational subspace $T \subset \mathbb{C}^n$, whose dimension we may take to be l.d.(z). The image of $\exp(T)$ is a subtorus $S \subset (\mathbb{C}^*)^n$, of the same dimension. $\square$

Claim 14.10   Suppose $(z, e^z) \in V - V'$. Then e^z lies in an atypical component of $W \cap S$.

Proof   We estimate tr.d.(z, e^z) below by SC and above by the intersection of V with the π pre image of $W \cap S$. Let A be the component of $W \cap T$ containing e^z. We find

$$\dim T \leq \text{tr.d.}(z, e^z) \leq d + \dim A$$
$$= \dim V - \dim W + \dim A < n - \dim W + \dim A.$$

Rearranging we see that

$$\dim A > \dim T + \dim W - n$$

and this proves the claim.

So if (z, e^z) lies in $V - V'$ then e^z satisfies one of finitely many multiplicative relations. Otherwise $(z, e^z) \in V'$, and we repeat the argument with its components V_i', putting $W_i' = \pi V_i'$ with generic fibre dimension d_i' outside $V_i'' \subset V'$, etc. $\square$

15 Zilber-Pink

Let X be a mixed Shimura variety, and $\mathcal{S}$ its collection of "special subvarieties". The following is essentially Zilber's formulation in Pink's setting. We define atypical components and Atyp(V) for $V \subset X$ exactly as previously.

Zilber-Pink Conjecture 15.1   Let $V \subset X$. Then Atyp(V) is a finite union.

It is natural to formulate this conjecture for $V/\mathbb{C}$, but it is again the version for $V/\mathbb{Q}$ that connects a Schanuel conjecture with its uniform version.

We consider the modular setting ($X = \mathbb{C}^n$).

Modular SC 15.2 (MSC) Let $z \in \mathbb{H}^n$ and let T be the smallest special subvariety of $\mathbb{H}^n$ containing z. Then

$$\mathrm{tr.d.}(z, j(z)) \geq \dim T.$$

UMSC 15.3 Let $V \subset \mathbb{C}^{2n}$ defined over $\mathbb{Q}$ and of dimension $\dim V < n$. There exist finitely many proper special subvarieties $T_1, \ldots, T_\ell$ such that if

$$(z, j(z)) \in V$$

then there exist T_i and $\gamma \in \mathrm{SL}_2(\mathbb{Z})^n$ such that $z \in T_i$. Equivalently, there are finitely many special subvarieties $S_i = j(T_i) \subset \mathbb{C}^n$ such that $j(z) \in S_i$ for some i.

Theorem 15.4 *MSC + MZP implies UMSC.*

Proof This is just the same as the proof of 14.9. Let V be given and define W, d, V' as before. Suppose $(z, j(z)) \in V$. Then $z \in T$ for some proper special T. Let $S = j(T)$, so $\dim S = \dim T$. Suppose $(z, j(z)) \in V - V'$. Let A be the component of $W \cap S$ containing $j(z)$. Then

$$\dim T \leq \mathrm{tr.d.}(z, j(z)) \leq d + \dim A$$
$$= \dim V - \dim W + \dim A < n - \dim W + \dim A.$$

So A is atypical and by MZP it is contained in an atypical component of one of finitely many proper specials S_i. Repeat for V'. □

16 Zilber-Pink and Ax-Schanuel

Other special cases of ZP have been successfully proved via o-minimality and point-counting (e.g. Masser-Zannier [36, 37, 38], Habegger-Pila [23], Bertrand-Masser-Pillay-Zannier [8], Orr [45], Bays-Habegger [6]) and in several of these one needs a suitable "Ax" type statement. In [23] it the modular analogue of "Ax-Logarithms", the algebraic independence of logarithms of algebraic functions.

Theorem 16.1 (Ax-Logarithms; Ax) *Suppose $C \subset (\mathbb{C}^*)^n$ is a curve. If $\log C$, locally on some disc, is contained in an algebraic hypersurface then C is contained in a proper weakly special subvariety.* □

Theorem 16.2 (Modular Ax-Logarithms; [23])   *Suppose $C \subset \mathbb{C}^n$ is a curve. If $j^{-1}(C)$, locally on some disc, is contained in an algebraic hypersurface then C is contained in a proper weakly special subvariety.*   □

This is proved using monodromy (not o-minimality). The required statements in Masser-Zannier are also proved using monodromy arguments.

In work in progress [24], Habegger-Pila show that "Weak Modular Ax-Schanuel" (as in §9) together with a suitable arithmetic statement that Galois orbits of certain atypical intersections are "large" implies, via o-minimality and point-counting, the full Zilber-Pink conjecture for $\mathbb{C}^n$.

Acknowledgements

I thank the participants at the LMS-EPSRC Short Course for their questions and comments which enabled me to clarify various issues in these notes. I am grateful to Alex Wilkie for pointing out many errors and inaccuracies in an earlier version. My thanks also to Boris Zilber for our numerous conversations on topics touched on here, which have much influenced my viewpoint. My work on these notes was partially supported by the EPSRC grant "O-minimality and Diophantine Geometry," EP/J019232/1.

References

[1] Y. André, *G-functions and geometry,* Aspects of Mathematics E13, Vieweg, Braunschweig, 1989.

[2] Y. André, *Une Introduction aux Motifs,* Panoramas et Sythèses, **17**, SMF, Paris 2004.

[3] J. Ax, On Schanuel's conjectures, *Annals* **93** (1971), 252–268.

[4] J. Ax, Some topics in differential algebraic geometry I: Analytic subgroups of algebraic groups, *Amer. J. Math.* **94** (1972), 1195–1204.

[5] A. Baker, *Transcendental Number Theory,* Cambridge University Press, 1974, 1979.

[6] M. Bays and P. Habegger, A note on divisible points on curves, arXiv:1301.5674 and *Trans. A. M. S.,* to appear.

[7] C. Bertolin, Périodes de 1-motifs et transcendance, *J. Number Th.* **97** (2002), 204–221.

[8] D. Bertrand, D. Masser, A. Pillay and U. Zannier, Relative Manin-Mumford for semi-abelian surfaces, manuscript, 2011/2013, arXiv:1307.1008.

[9] D. Bertrand and A. Pillay, A Lindemann-Weierstrass theorem for semi-abelian varieties over function fields, *J. Amer. Math. Soc.* **23** (2010), 491–533.

[10] D. Bertrand and W. Zudilin, Derivatives of Siegel modular forms and exponential functions, *Izvestiya Math.* **65** (2001), 659–671.

[11] E. Bombieri, P. Habegger, D. Masser and U. Zannier, A note on Maurin's theorem, *Rend. Lincei. Mat. Appl.* **21** (2010), 251–260.

[12] E. Bombieri, D. Masser and U. Zannier, Intersecting a curve with algebraic subgroups of multiplicative groups, *IMRN* **20** (1999), 1119–1140.

[13] E. Bombieri, D. Masser and U. Zannier, Anomalous subvarieties – structure theorems and applications, *IMRN* **19** (2007), 33 pages.

[14] E. Bombieri, D. Masser and U. Zannier, Intersecting a plane with algebraic subgroups of multiplicative groups, *Ann. Scuola Norm. Pisa Cl. Sci (5)* **7** (2008), 51–80.

[15] E. Bombieri, D. Masser and U. Zannier, On unlikely intersections of complex varieties with tori, *Acta Arithmetica* **133** (2008), 309–323.

[16] C. Daw, this volume.

[17] F. Diamond and J. Shurman, *A First Course in Modular Forms,* Graduate Texts in Mathematics **228**, Springer, New York, 2005.

[18] G. Diaz, Transcendence et indépendance algébriques: liens entre les points de vue elliptique et modulaire, *Ramanujan J.* **4** (2000), 157–199.

[19] G. van der Geer, Siegel modular forms and their applications, in *The 1-2-3 of Modular Forms,* Universitext, Springer, Berlin, 2008.

[20] P. Habegger, Intersecting subvarieties of $\mathbf{G}_m^n$ with algebraic subgroups, *Math. Annalen* **342** (2008), 449–466.

[21] P. Habegger, On the bounded height conjecture, *IMRN* **2009**, 860–886.

[22] P. Habegger, Torsion points on elliptic curves in Weierstrass form, *Ann. Sc. Norm. Sup. Pisa Cl. Sci. (5)* **12** (2013), 687–715.

[23] P. Habegger and J. Pila, Some unlikely intersections beyond André–Oort, *Compositio* **148** (2012), 1–27.

[24] P. Habegger and J. Pila, O-minimality and certain atypical intersections, arXiv:1409.0771.

[25] J. Kirby, The theory of the exponential differential equations of semiabelian varieties, *Selecta Math.* **15** (2009), 445–486.

[26] J. Kirby, Variants of Schanuel's conjecture, manuscript, 2007.

[27] J. Kirby and B. Zilber, The uniform Schanuel conjecture over the real numbers, *Bulletin London Math Soc.* **38** (2006), 568–570.

[28] J. Kirby and B. Zilber, Exponentially closed fields and the Conjecture on Intersections with Tori, *Ann. Pure Appl. Logic* **165** (2014), 1680–1706.

[29] B. Klingler and A. Yafaev, The André-Oort conjecture, *Annals* **180** (2014), 867–925.

[30] B. Klingler, E. Ullmo and A. Yafaev, The hyperbolic Ax-Lindemann-Weierstrass conjecture, 2013 manuscript.

[31] S. Lang, *Introduction to transcendental numbers,* Addison-Wesley, Reading, 1966.

[32] S. Lang, *Algebra,* revised third edition, Graduate Texts in Mathematics **211**, Springer, New York, 2002. (Or earlier editions published by Addison-Wesley.)

[33] S. Lang, *Elliptic functions,* second edition, Graduate Texts in Mathematics **112**, Springer, New York, 1987.

[34] K. Mahler, On algebraic differential equations satisfied by automorphic functions, *J. Austral. Math. Soc.* **10** (1969), 445–450.

[35] D. Masser, Heights, Transcendence, and Linear Independence on Commutative Group Varieties, Lecture Notes in Mathematics **1819**, Amoroso and Zannier, eds, 1-51, Springer-Verlag, Berlin, 2003.

[36] D. Masser and U. Zannier, Torsion anomalous points and families of elliptic curves, *C. R. Math. Acad. Sci. Paris* **346** (2008), 491–494.

[37] D. Masser and U. Zannier, Torsion anomalous points and families of elliptic curves, *Amer. J. Math.* **132** (2010), 1677–1691.

[38] D. Masser and U. Zannier, Torsion points on families of squares of elliptic curves, *Math. Ann.* **352** (2012), 453–484.

[39] G. Maurin, Courbes algébriques et équations multiplicatives, *Math. Annalen* **341** (2008), 789–824.

[40] G. Maurin, Équations multiplicatives sur les sous-variétés des tores, *IMRN* **2011**.

[41] B. J. J. Moonen, Linearity properties of Shimura varieties, I, *J. Alg. Geom.* **7** (1998), 539–567.

[42] D. Mumford, *Algebraic Geometry I: Complex Projective Varieties*, Grundlehren der math. Wiss. **221**, corrected second printing, Springer, Berlin, 1976.

[43] Yu. V. Nesterenko, *Algebraic Independence*, TIFR Mumbai/Narosa, 2009.

[44] F. Oort, Canonical lifts and dense sets of CM points, *Arithmetic Geometry, Cortona, 1994*, 228–234, F. Catanese, editor, Symposia. Math., XXXVII, CUP, 1997.

[45] M. Orr, Families of abelian varieties with many isogenous fibres, *Crelle,* to appear.

[46] Y. Peterzil and S. Starchenko, Uniform definability of the Weierstrass $\wp$ functions and generalized tori of dimension one, *Selecta Math. N. S.* **10** (2004), 525–550.

[47] Y. Peterzil and S. Starchenko, Definability of restricted theta functions and families of abelian varieties, *Duke Math. J.* **162** (2013), 731–765.

[48] J. Pila, O-minimality and the André-Oort conjecture for $\mathbb{C}^n$, *Annals* **173** (2011), 1779–1840.

[49] J. Pila, Modular Ax-Lindemann-Weierstrass with derivatives, *Notre Dame J. Formal Logic* **54** (2013), 553–565 (Oléron proceedings).

[50] J. Pila, Special point problems with elliptic modular surfaces, *Mathematika* **60** (2014), 1–31.

[51] J. Pila and J. Tsimerman, The André-Oort conjecture for the moduli space of abelian surfaces, *Compositio* **149** (2013), 204–216.

[52] J. Pila and J. Tsimerman, Ax-Lindemann for $\mathcal{A}_g$, *Annals* **179** (2014), 659–681.

[53] J. Pila and A. J. Wilkie, The rational points of a definable set, *DMJ* **133** (2006), 591–616.

[54] J. Pila and U. Zannier, Rational points in periodic analytic sets and the Manin-Mumford conjecture, *Rend. Lincei Mat. Appl.* **19** (2008), 149–162.

[55] R. Pink, A combination of the conjectures of Mordell-Lang and André-Oort, *Geometric methods in algebra and number theory,* F. Bogomolov, Y. Tschinkel, editors, pp. 251–282, Prog. Math. **253**, Birkhauser, Boston MA, 2005.

[56] R. Pink, A common generalization of the conjectures of André-Oort, Manin-Mumford, and Mordell-Lang, 2005 preprint, available from the author's webpage.

[57] J.-P. Serre, *A Course in Arithmetic,* Graduate Texts in Mathematics **7**, Springer, New York, 1973.

[58] J. Tsimerman, Brauer-Siegel for arithmetic tori and lower bounds for Galois orbits of special points, *J. Amer. Math. Soc.* **25** (2012), 1091–1117.

[59] J. Tsimerman, Ax-Schanuel and O-minimality, this volume.

[60] E. Ullmo, Quelques applications du théorème de Ax-Lindemann hyperbolique, *Compositio* **150** (2014), 175–190.

[61] E. Ullmo and A. Yafaev, Galois orbits and equidistribution of special subvarieties: towards the André-Ort conjecture, *Annals* **180** (2014), 823–865.

[62] E. Ullmo and A. Yafaev, A characterisation of special subvarieties, *Mathematika* **57** (2011), 833–842.

[63] E. Ullmo and A. Yafaev, Nombres de classes des tores de multiplication complexe et bornes inférieures pour orbites Galoisiennes de points spéciaux, arXiv:1209.0942.

[64] E. Ullmo and A. Yafaev, Hyperbolic Ax-Lindemann in the Cocompact case, *Duke Math. J.*, **163** (2014), 433–463.

[65] M. Waldschmidt, *Diophantine Approximation on Linear Algebraic Groups*, Grund- leheren der math. Wiss **326**, Springer, Berlin, 2000.

[66] M. Waldschmidt, *Some Consequences of Schanuel's Conjecture*, Talk slides available from the author's website at www.math.jussieu.fr/ miw/

[67] D. Zagier, Elliptic modular functions and their applications, in *The 1-2-3 of Modular Forms*, Universitext, Springer, Berlin, 2008.

[68] U. Zannier, *Some problems of unlikely intersections in arithmetic and geometry*, with appendices by D. Masser, *Annals of Mathematics Studies* **181**, Princeton University Press, 2012.

[69] B. Zilber, Pseudo-exponentiation on algebraically closed fields of characteristic zero, *Ann. Pure Appl. Logic* **132** (2005), 67–95.

[70] B. Zilber, Intersecting varieties with tori, 2001 preprint incorporated into [71].

[71] B. Zilber, Exponential sums equations and the Schanuel conjecture, *J. London Math. Soc. (2)* **65** (2002), 27–44.

[72] B. Zilber, *On transcendental number theory, classical analytic functions and Diophantine geometry*, Talk slides available from http://people.maths.ox.ac.uk/zilber/

[73] B. Zilber, Model theory of special subvarieties and Schanuel-type conjectures, 2013 preprint available from http://people.maths.ox.ac.uk/zilber/

4

Introduction to abelian varieties and the Ax–Lindemann–Weierstrass theorem

Martin Orr

1 Introduction

This paper surveys some aspects of the theory of abelian varieties relevant to the Pila–Zannier proof of the Manin–Mumford conjecture and to the André–Oort conjecture. An abelian variety is a complete algebraic variety with a group law. The geometry of abelian varieties is tightly constrained and well-behaved, and they are important tools in algebraic geometry. Abelian varieties defined over number fields pose interesting arithmetic problems, for example concerning their rational points and associated Galois representations.

The paper is in three parts:

(1) an introduction to abelian varieties;
(2) an outline of moduli spaces of principally polarised abelian varieties, which are the fundamental examples of Shimura varieties;
(3) a detailed proof of the Ax–Lindemann–Weierstrass theorem for abelian varieties, following a method using o-minimal geometry due to Pila, Ullmo and Yafaev.

The first part assumes only an elementary knowledge of algebraic varieties and complex analytic geometry. The second part makes heavier use of algebraic geometry, but still at the level of varieties, and a little algebraic number theory. Like the first part, the algebraic geometry in the third part is elementary; the third part also assumes familiarity with the concept of semialgebraic sets, and uses cell decomposition for semialgebraic sets and the Pila–Wilkie theorem as black boxes. The second and third parts are independent of each other, so the reader interested primarily in the Ax–Lindemann–Weierstrass theorem may skip the second part (sections 4 to 6).

O-Minimality and Diophantine Geometry, ed. G. O. Jones and A. J. Wilkie. Published by Cambridge University Press. © Cambridge University Press 2015.

In the first part of the paper (sections 2 and 3), we introduce abelian varieties over fields of characteristic zero, and especially over the complex numbers. The theory of abelian varieties over fields of positive characteristic introduces additional complications which we will not discuss. Our choice of topics is driven by Pila and Zannier's proof of the Manin–Mumford conjecture using o-minimal geometry. We will not discuss the Manin–Mumford conjecture or its proofs directly in this paper; aspects of the proof, and its generalisation to Shimura varieties, are discussed in other papers in this volume.

In this part of the article, we omit proofs of most statements. Most of the material is covered in Milne's online notes [Mil08] (which are a revised version of [Mil86]) and in Birkenhake and Lange's book [BL92] (over the complex numbers only). Another standard reference is Mumford's book [Mum70], which contains some proofs which are omitted in [Mil08] but deals only with algebraically closed fields (of any characteristic) and is a more difficult read. We have not attempted to cite original sources for theorems, but simply refer to whichever of these books offers the most convenient proof for each result.

The second part of the paper (sections 4 to 6) outlines the definition and analytic construction of the moduli spaces of principally polarised abelian varieties. These are algebraic varieties whose points parametrise abelian varieties. The purpose of this part is to give some concrete examples of Shimura varieties to complement the more abstract discussion in Daw's paper on the André–Oort conjecture. These sections require a higher level of algebraic geometry than the rest of the paper, and section 6 on complex multiplication also uses algebraic number fields.

Again in these sections we omit proofs. Proofs for section 4 can be found in the same books mentioned above. For section 5, see chapter 7 of [Mum65]; there is also a quick sketch in [Mil08] III, section 7. Details of the theory of complex multiplication (section 6) may be found in the online notes [Mil06]; for a gentler introduction, you may restrict your attention to the case of elliptic curves which is discussed in [Sil94] chapter 2.

In the third part of the paper we discuss the Ax–Lindemann–Weierstrass theorem for abelian varieties. This theorem has its roots in transcendence theory and is one of the ingredients in the Pila–Zannier approach to the Manin–Mumford conjecture.

Theorem 1.1 *Let*

- *A be an abelian variety of dimension g over $\mathbb{C}$,*
- *$\pi : \mathbb{C}^g \to A$ be the exponential map,*
- *V be a complex algebraic subvariety of A, and*

- *Y be a maximal irreducible complex algebraic subvariety contained in* $\pi^{-1}(V)$.

Then $\pi(Y)$ is a translate of an abelian subvariety of A.

Several proofs of this theorem are known. A proof using the Pila–Wilkie theorem was given by Pila in [Pil11]. This proof forms the basis for subsequent proofs of analogous theorems for Shimura varieties. The proof we will describe is based on Pila's ideas, but incorporating part of Ullmo and Yafaev's proof of the Ax–Lindemann–Weierstrass theorem for compact Shimura varieties [UY13] in order to simplify the proof.

2 Abelian varieties

In this section we will define abelian varieties and their morphisms and state their basic properties, and those of their torsion points. We work over an arbitrary base field, although some of the theorems will include a condition on the characteristic and our main interest is in base fields of characteristic zero.

2.1 Definitions and basic properties

Definition   An **abelian variety** is a complete group variety.

Let us define the terms which appear in the above definition.

Definition   A **group variety** is an algebraic variety G together with morphisms of varieties $m\colon G \times G \to G$ (multiplication) and $i\colon G \to G$ (inverse) and a point $e \in G$ (the identity element) which satisfy the axioms for a group.

When we say that a group variety is defined over a field k, we mean that the variety G, the morphisms m and i and the point e are all defined over k.

Definition   An algebraic variety X over an algebraically closed field is **complete** if, for every variety Y, the projection $X \times Y \to Y$ is a closed map with respect to the Zariski topologies on $X \times Y$ and Y. (If the base field is not algebraically closed, then we should make the same definition using schemes instead of varieties.)

The property of completeness is the analogue in algebraic geometry of compactness in topology. The fundamental examples of complete varieties are

projective varieties, that is, closed subvarieties of projective space. In general, it is not true that all complete varieties are projective. In the case of abelian varieties, it turns out that all abelian varieties are projective but this is a far-from-trivial theorem. Indeed, Weil did not know this theorem when he first developed the theory of abelian varieties.

Theorem 2.1 ([Mil08] I, section 6)   *Every abelian variety is projective.*

Another fundamental geometric property of abelian varieties, which they share with all group varieties, is that they are smooth.

The twin conditions of being complete and having a group law imply that abelian varieties are topologically and group-theoretically boring; in this they differ from compact Lie groups. At the heart of this is the following rigidity result, which follows from the definition of completeness together with the fact that every morphism from a complete variety to an affine variety is constant.

Theorem 2.2 ([Mil08] I, Theorem 1.1)   *Let X, Y and Z be complete varieties and $f: X \times Y \to Z$ a morphism. Suppose that there exist points $x \in X$ and $y \in Y$ such that the restrictions $f_{|\{x\} \times Y}$ and $f_{|X \times \{y\}}$ are constant.*
Then f is constant.

By applying Theorem 2.2 to the conjugator map

$$(x, y) \mapsto xyx^{-1}y^{-1} : A \times A \to A$$

we deduce the following corollary.

Corollary 2.3   *The group law on an abelian variety is commutative.*

As a result of this, we shall henceforth write the group law on an abelian variety additively.

We can also show that the group law on an abelian variety over an algebraically closed field of characteristic zero is divisible. Hence apart from describing the torsion, which we shall do later, and rationality questions over non-algebraically closed fields, there is nothing to say about the group theory of abelian varieties.

2.2 Elliptic curves

The simplest examples of abelian varieties are **elliptic curves**, which by definition are abelian varieties of dimension 1.

Elliptic curves are commonly described as curves of the form

$$y^2 = x^3 + ax + b$$

for constants a and b satisfying $4a^3 + 27b^2 \neq 0$, or more correctly as the closures of such curves in $\mathbb{P}^2$. The closure of such a curve consists of the affine curve together with the single point $[0 : 1 : 0]$ at infinity. The condition $4a^3 + 27b^2 \neq 0$, or equivalently that the cubic $x^3 + ax + b$ has distinct roots, ensures that the curve is smooth.

Given such a curve, we can define a group law using the classical chord-and-tangent construction, with the point at infinity as the identity element. Thus we get an abelian variety of dimension 1.

Conversely, every abelian variety of dimension 1 is isomorphic to one of the above form.

2.3 Morphisms and isogenies

A **morphism** of abelian varieties is a morphism of algebraic varieties which is also a group homomorphism.

An **isogeny** is a morphism of abelian varieties which is surjective and has finite kernel. Note that the domain and codomain of an isogeny always have the same dimension.

The **degree** of an isogeny $f \colon A \to B$ is its degree as a morphism of algebraic varieties, or in other words the degree of the associated extension of fields of rational functions $[k(A) : k(B)]$. Over an algebraically closed field of characteristic zero, this is the same as the number of points in the kernel of the isogeny.

Examples of isogenies are the multiplication-by-n maps from an abelian variety to itself. We will see later that this fact is easy to prove analytically when A is defined over $\mathbb{C}$. A general proof is much harder.

Proposition 2.4 ([Mil08] I, Theorem 7.2)   *Let A be an abelian variety and N a non-zero integer. Let $[N] \colon A \to A$ be the morphism which sends x to Nx in the group law. Then $[N]$ is an isogeny of degree N^{2g}.*

We say that two abelian varieties A and B are **isogenous** if there exists an isogeny $A \to B$. The relationship of being isogenous is an equivalence relation – the hard part of this to prove is that if there is an isogeny $A \to B$ then there is also an isogeny $B \to A$.

2.4 Abelian subvarieties

Let A be an abelian variety. An **abelian subvariety** of A is an algebraic subvariety which is also a subgroup.

Given any abelian subvariety $B \subset A$, we can construct a quotient abelian variety A/B. It is not necessarily true that A is isomorphic to the direct product $B \times A/B$, only that A is isogenous to $B \times A/B$. If we attempt to decompose A as an internal direct product, we get the following theorem called the Poincaré Reducibility Theorem.

Theorem 2.5 ([Mil08] proof of I, Proposition 10.1) *Let A be an abelian variety and $B \subset A$ an abelian subvariety. Then there exists an abelian subvariety $C \subset A$ such that B and C together generate A and $B \cap C$ is finite (but $B \cap C$ might not be trivial).*

A **simple abelian variety** is an abelian variety A whose only abelian subvarieties are A itself and $\{0\}$. The Poincaré Reducibility Theorem implies that every abelian variety is isogenous (but not necessarily isomorphic) to a direct product of simple abelian varieties.

2.5 Torsion points

Let A be an abelian variety of dimension g and let N be a positive integer. We write $A[N]$ for the set of N-torsion points, that is, points $x \in A$ such that $Nx = 0$ in the group law on A.

When talking about N-torsion points, we should assume that the characteristic of the base field does not divide N – otherwise the kernel of the multiplication map $[N]$ is a non-reduced group scheme and not just a set of points. If we make this assumption on the characteristic, and also suppose that the base field is algebraically closed, then the following description of the group of N-torsion points is equivalent to Proposition 2.4.

Proposition 2.6 *Let A be an abelian variety of dimension g defined over an algebraically closed field whose characteristic does not divide N. Then $A[N]$ is isomorphic to $(\mathbb{Z}/N\mathbb{Z})^{2g}$ as a group.*

Suppose that A is defined over an arbitrary field k whose characteristic does not divide N. If P is a $\bar{k}$-point of $A[N]$, then the $\mathrm{Gal}(\bar{k}/k)$-conjugates of P are also N-torsion points of A. Hence $\mathrm{Gal}(\bar{k}/k)$ acts on $A[N](\bar{k})$. These Galois actions are much studied, especially in the cases where k is a number field, local field or function field.

Let $P \in A(\bar{k})$ be a torsion point of order N. Then the degree $[k(P) : k]$ of the field of definition of P is bounded above by $\#A[N](\bar{k}) = N^{2g}$. We can improve this upper bound slightly by excluding the points of $A[N](\bar{k})$ whose order is smaller than N, but this only gives lower order terms in the bound.

When k is a number field, we can also prove a lower bound for $[k(P) : k]$. This generalises the bound for torsion points on elliptic curves discussed in Habegger's paper in this volume.

Theorem 2.7 ([Mas84]) *Let A be an abelian variety of dimension g defined over a number field k. There exist effective constants c, depending on A and k, and ρ, depending only on g, such that for all torsion points $P \in A(\bar{k})$ of order N,*

$$[k(P) : k] \geq cN^{\rho}.$$

3 Complex tori

There is a very simple analytic description of abelian varieties over the complex numbers, as complex tori. This analytic description is essential to the Pila–Zannier proof of the Manin–Mumford conjecture, and is needed for the statement of the Ax–Lindemann–Weierstrass theorem.

Let V be a finite-dimensional complex vector space, which we consider as a group under addition. A **lattice** in V is a discrete subgroup $\Lambda \subset V$ such that the quotient V/Λ is compact. Observe that if Λ is a lattice, then Λ is isomorphic as a group to $\mathbb{Z}^{2g}$ (where $g = \dim_{\mathbb{C}} V$) and that the $\mathbb{R}$-span of Λ is V.

A complex manifold of the form V/Λ, where V is a complex vector space and $\Lambda \subset V$ is a lattice, is called a **complex torus**. Note that the word "torus" is confusingly over-used in the world of group varieties. In this case, it is used because V/Λ is diffeomorphic to $\mathbb{R}^{2g}/\mathbb{Z}^{2g} \cong (S^1)^{2g}$; so in particular, if $g = 1$ then V/Λ is diffeomorphic to the classical torus $\mathbb{R}^2/\mathbb{Z}^2$.

Theorem 3.1 *Every abelian variety over $\mathbb{C}$ is isomorphic as a complex Lie group to a complex torus.*

Proof Let A be an abelian variety over $\mathbb{C}$ and let V be its tangent space at the identity. Because $A(\mathbb{C})$ is a complex Lie group, there is a holomorphic exponential map $\exp \colon V \to A(\mathbb{C})$. Because A is commutative, $\exp$ is a surjective group homomorphism. Its kernel Λ is a lattice, and $A(\mathbb{C}) \cong V/\Lambda$. $\square$

Note that the converse is false – not every complex torus is isomorphic to an abelian variety. This will be discussed in section 4.

3.1 Morphisms of complex tori

Let A and A' be complex abelian varieties, isomorphic to the complex tori V/Λ and V'/Λ' respectively. If $f\colon A \to A'$ is a morphism of abelian varieties, then it lifts to a $\mathbb{C}$-linear map $\tilde{f}\colon V \to V'$ such that $\tilde{f}(\Lambda) \subset \Lambda'$.

Conversely, any $\mathbb{C}$-linear map $V \to V'$ which maps Λ into Λ' descends to a holomorphic map $V/\Lambda \to V'/\Lambda'$, and by Chow's theorem this is a morphism of algebraic varieties $A \to A'$. Hence there is a canonical bijection

$$\left\{\begin{array}{c} \text{morphisms of abelian} \\ \text{varieties } A \to A' \end{array}\right\} \longleftrightarrow \left\{\begin{array}{c} \mathbb{C}\text{-linear maps } V \to V' \\ \text{mapping } \Lambda \text{ into } \Lambda' \end{array}\right\}$$

3.2 Abelian subvarieties of complex tori

Let A be a complex abelian variety isomorphic to V/Λ. Suppose that A' is an abelian subvariety of A. Then $A' = V'/\Lambda'$ where V' is some complex vector subspace of V and $\Lambda' = V' \cap \Lambda$.

Conversely, let V' be a complex vector subspace of V. We say that V' is a **full subspace** if $V' \cap \Lambda$ is a lattice in V'. Note that most subspaces are not full – indeed for a generic subspace V', $V' \cap \Lambda = \{0\}$.

Clearly V' being full is a necessary condition for $V'/(V' \cap \Lambda)$ to be an abelian variety. If V' is full, then Chow's theorem implies that $V'/(V' \cap \Lambda)$ is in fact an abelian subvariety of A.

Hence we get a canonical bijection

$$\{\text{abelian subvarieties of } A\} \longleftrightarrow \{\text{full } \mathbb{C}\text{-vector subspaces of } (V, \Lambda)\}.$$

3.3 Torsion points in complex tori

The N-torsion points of the complex torus V/Λ are $\frac{1}{N}\Lambda/\Lambda$. Since $\Lambda \cong \mathbb{Z}^{2g}$, where $g = \dim_{\mathbb{C}} V$, there is a group isomorphism $\frac{1}{N}\Lambda/\Lambda \cong (\mathbb{Z}/N\mathbb{Z})^{2g}$. This gives a quick proof of Proposition 2.4 for abelian varieties defined over $\mathbb{C}$.

4 Riemann forms and polarisations

In the previous section, we saw that every complex abelian variety is a complex torus. The question remains of which complex tori are abelian varieties – in other words (given Theorem 2.1), when can the complex manifold V/Λ be embedded as a closed subvariety of some projective space? The answer to this question is provided by Riemann forms.

In this section, we will define Riemann forms, then briefly discuss dual abelian varieties and polarisations. Riemann forms are defined complex analytically; polarisations provide a substitute which can be generalised to arbitrary base fields. We will use polarisations as a technical tool when defining moduli spaces.

4.1 Riemann forms

The definition of Riemann forms is rather opaque; they are simply what is needed to answer the question of which complex tori are abelian varieties.

Let V be a complex vector space and Λ a lattice in V. A **Riemann form** on (V, Λ) is a symplectic bilinear form $\psi \colon \Lambda \times \Lambda \to \mathbb{Z}$ such that:

(1) $\psi_{\mathbb{R}}(iu, iv) = \psi_{\mathbb{R}}(u, v)$ for all $u, v \in V$; and
(2) $\psi_{\mathbb{R}}(iv, v) > 0$ for all $v \in V - \{0\}$.

Here, $\psi_{\mathbb{R}} \colon V \times V \to \mathbb{R}$ means the $\mathbb{R}$-bilinear extension of ψ (note that this is not $\mathbb{C}$-bilinear – indeed it cannot be $\mathbb{C}$-bilinear because its image is contained in $\mathbb{R}$). Condition (2) implies that ψ is non-degenerate.

There is a second perspective on Riemann forms, using Hermitian forms on V instead of symplectic forms on Λ.

Lemma 4.1   *The map sending a Hermitian form to its imaginary part is a bijection*

$$\left\{ \begin{array}{c} \textit{positive definite Hermitian forms} \\ H \colon V \times V \to \mathbb{C} \textit{ s.t. } \operatorname{Im} H(\Lambda \times \Lambda) \subset \mathbb{Z} \end{array} \right\} \to \left\{ \begin{array}{c} \textit{Riemann forms} \\ \textit{on } (V, \Lambda) \end{array} \right\}$$

whose inverse sends a Riemann form ψ to the Hermitian form

$$H(u, v) = \psi_{\mathbb{R}}(iu, v) + i\psi_{\mathbb{R}}(u, v).$$

The importance of Riemann forms is due to the following theorem.

Theorem 4.2 ([Mum70] section 3, Corollary)   *A complex torus V/Λ is isomorphic to an abelian variety if and only if there exists a Riemann form on (V, Λ).*

In fact this theorem can be strengthened as follows.

Theorem 4.3 *There is a canonical bijection between the set of Riemann forms on (V, Λ) and the set of homological equivalence classes of ample line bundles on V/Λ.*

We can interpret the lattice Λ, together with the action of $\mathbb{C}$ on $V = \Lambda \otimes_{\mathbb{Z}} \mathbb{R}$, as a Hodge structure of type $\{(-1, 0), (0, -1)\}$ – for more on this, see Daw's paper. A Riemann form $\psi \colon \Lambda \times \Lambda \to \mathbb{Z}$ is the same as what is known as a polarisation in Hodge theory. It might seem a little confusing that we are about to define a different type of object called a polarisation in the context of abelian varieties; however, because there is a canonical bijection between Riemann forms and polarisations as we will define them below they are really two different ways of looking at the same thing.

4.2 Polarisations over $\mathbb{C}$

If $A \cong V/\Lambda$ is an abelian variety over $\mathbb{C}$, then for each Riemann form on (V, Λ) we can construct an isogeny from A to the so-called dual abelian variety. The isogenies constructed in this way are called polarisations. Because polarisations are objects of algebraic rather than analytic geometry, they can be generalised to abelian varieties over arbitrary base fields.

Let V be a complex vector space and Λ a lattice in V. Let $\bar{V}^{\vee}$ be the conjugate-dual space of V, that is

$$\bar{V}^{\vee} = \{f \colon V \to \mathbb{C} \mid f \text{ is additive and } f(av) = \bar{a}f(v) \text{ for all } a \in \mathbb{C}, v \in V\}.$$

Let $\bar{\Lambda}^{\vee}$ be the following lattice in $\bar{V}^{\vee}$

$$\bar{\Lambda}^{\vee} = \{f \colon V \to \mathbb{C} \mid \operatorname{Im} f(\Lambda) \subset \mathbb{Z}\}$$

(where Im means the imaginary part).

Suppose that V/Λ is isomorphic to an abelian variety A. Then there exists a Riemann form $\psi \colon \Lambda \times \Lambda \to \mathbb{Z}$. We can use ψ to construct a Riemann form on $(\bar{V}^{\vee}, \bar{\Lambda}^{\vee})$. Hence $\bar{V}^{\vee}/\bar{\Lambda}^{\vee}$ is isomorphic to an abelian variety, which we denote $A^{\vee}$ and call the **dual abelian variety** of A.

Let $H \colon V \times V \to \mathbb{C}$ be the Hermitian form associated with the Riemann form ψ as in Lemma 4.1. Then

$$v \mapsto H(v, -)$$

is a $\mathbb{C}$-linear isomorphism $V \to \bar{V}^{\vee}$ which maps Λ into $\bar{\Lambda}^{\vee}$, and hence induces a morphism of abelian varieties $\lambda \colon A \to A^{\vee}$. The fact that ψ is non-degenerate implies that λ is an isogeny.

Any isogeny $A \to A^\vee$ which is induced by a Riemann form according to the above recipe is called a **polarisation** of A.

4.3 Polarisations over arbitrary fields

The details of how to define dual abelian varieties and polarisations over arbitrary fields are beyond the scope of this paper. We merely remark that, if A is an abelian variety defined over a field k, then it is possible to define the dual abelian variety $A^\vee$ over k, and that a polarisation of A is an isogeny $A \to A^\vee$ which is induced according to a certain recipe by an ample line bundle on A (Theorem 4.3 implies that this is equivalent to the previous definition when $k = \mathbb{C}$). Because every abelian variety is projective, there always exists at least one polarisation on A.

4.4 Principal polarisations

A polarisation is said to be **principal** if it has degree 1 as an isogeny, or in other words, if it is an isomorphism between A and $A^\vee$. Not all abelian varieties possess principal polarisations, but the following results mean that it is often possible to reduce a question about general abelian varieties to a question about abelian varieties with principal polarisations:

Proposition 4.4 ([Mil08] I, Corollary 13.10, Theorem 13.12)

(1) Over a field of characteristic zero, every abelian variety is isogenous to an abelian variety with a principal polarisation.

(2) ("Zarhin's Trick") Over any base field, if A is any abelian variety, then $(A \times A^\vee)^4$ has a principal polarisation.

4.5 Polarisations for elliptic curves

In one dimension, every lattice $\Lambda \subset \mathbb{C}$ has a Riemann form and so $\mathbb{C}/\Lambda$ is an elliptic curve. To write down the Riemann form explicitly, rescale Λ so that it has a $\mathbb{Z}$-basis $\{1, \tau\}$ with $\text{Im}\, \tau > 0$. Then

$$(u, v) \mapsto \frac{\text{Im}(u\bar{v})}{\text{Im}\, \tau}$$

is a Riemann form on $(\mathbb{C}, \Lambda)$.

The above Riemann form defines a principal polarisation of the corresponding elliptic curve E. In fact this is the unique principal polarisation of E so it gives a canonical isomorphism $E \to E^\vee$. Hence it is rarely necessary to explicitly talk about polarisations and dual varieties when working with elliptic curves.

5 The moduli space of principally polarised abelian varieties

The moduli space of principally polarised abelian varieties of dimension g is an algebraic variety $\mathcal{A}_g$ equipped with a bijection between its points (over an algebraically closed field) and the set of isomorphism classes of principally polarised abelian varieties of dimension g, satisfying a sort of weakened universal property which we shall discuss below.

We will begin by explaining how this universal property is made precise, giving a definition of the moduli space as a variety over any perfect field. We then outline Siegel's analytic construction of the moduli space over the complex numbers. This is a prototype for the definition of a Shimura variety as a quotient of a Hermitian symmetric domain by the action of an arithmetic group. Historically, the analytic construction came first, and Satake and Baily proved that the complex analytic moduli space is an algebraic variety defined over $\mathbb{Q}$ before the algebraic theory was developed by Mumford.

5.1 Principally polarised abelian varieties

There is no "moduli space of abelian varieties of dimension g": in order to get a moduli space it is necessary to include polarisations.

A **principally polarised abelian variety** is a pair (A, λ) where A is an abelian variety and $\lambda \colon A \to A^\vee$ is a principal polarisation. An **isomorphism** between principally polarised abelian varieties (A, λ) and (B, μ) is an isomorphism $f \colon A \to B$ of abelian varieties such that

$$\lambda = f^\vee \circ \mu \circ f$$

where $f^\vee \colon B^\vee \to A^\vee$ is the dual morphism of f.

Theorem 5.1 ([Mil08] I, Theorem 15.1) *Let A be an abelian variety. There are finitely many isomorphism classes of principally polarised abelian varieties (A, λ).*

Note that the theorem does not assert that A has finitely many principal polarisations, because there may be distinct principal polarisations λ and λ' such that (A, λ) and (A, λ') are isomorphic as principally polarised abelian varieties.

5.2 Universal property definition of the moduli space

The idea behind the universal property definition of the moduli space $\mathcal{A}_g$ is that it should be the base variety for a "universal principally polarised family of abelian varieties". The problem is that no such universal family exists. The best we can do is to require that $\mathcal{A}_g$ satisfies the property that it would satisfy if a universal family did exist (property (a) below). Since this is not enough to determine $\mathcal{A}_g$ up to isomorphism, we add an additional condition that $\mathcal{A}_g$ is universal among varieties satisfying (a). A variety satisfying properties with the form of (a) and (b) is called a **coarse moduli space**.

We will first work over an algebraically closed field k.

A **family of abelian varieties** is a morphism of algebraic varieties $\pi : A \to S$ together with morphisms $m : A \times A \to A$, $i : A \to A$ and $e : S \to A$ such that:

(i) π is proper; and

(ii) for each point $s \in S(k)$, the fibre A_s together with the restrictions $m_{|A_s \times A_s}$ and $i_{|A_s}$ and the point $e(s)$ form an abelian variety.

It is possible to define a notion of principally polarised family of abelian varieties.

The **moduli space of principally polarised abelian varieties of dimension** g over k is an algebraic variety $\mathcal{A}_g$ defined over k equipped with a bijection

$$J_g : \left\{ \begin{array}{l} \text{isomorphism classes of principally polarised} \\ \text{abelian varieties of dimension } g \text{ defined over } k \end{array} \right\} \longrightarrow \mathcal{A}_g(k)$$

satisfying the following conditions:

(a) for every algebraic variety T defined over k and every principally polarised family of abelian varieties $B \to T$ of relative dimension g, if φ denotes the map $T(k) \to \mathcal{A}_g(k)$ which sends $t \in T(k)$ to $J_g(B_t) \in \mathcal{A}_g(k)$, then φ is a morphism of algebraic varieties;

(b) for every algebraic variety $\mathcal{A}_g'$ over k equipped with a bijection J_g' as above, if $(\mathcal{A}_g', J_g')$ satisfies (a), then $J_g' \circ J_g^{-1} : \mathcal{A}_g(k) \to \mathcal{A}_g'(k)$ is a morphism of algebraic varieties.

These properties are sufficient to characterise the variety $\mathcal{A}_g$, together with the bijection J_g, up to unique isomorphism.

Over a non-algebraically closed perfect field k, we define the moduli space of principally polarised abelian varieties to be an algebraic variety defined over k together with a bijection

$$\left\{ \begin{array}{c} \text{isomorphism classes of principally polarised} \\ \text{abelian varieties of dimension } g \text{ defined over } \bar{k} \end{array} \right\} \longrightarrow \mathcal{A}_g(\bar{k})$$

satisfying the properties (a) and (b) over the algebraic closure $\bar{k}$, and such that the bijection commutes with the natural actions of $\mathrm{Gal}(\bar{k}/k)$ on either side. When k is non-algebraically closed, there is still a map from isomorphism classes of principally polarised abelian varieties defined over k to the k-points of $\mathcal{A}_g$, but it is usually neither injective nor surjective.

Theorem 5.2 ([Mum65] chapter 7) *The moduli space of principally polarised abelian varieties of dimension g exists over any base field k.*

5.3 Analytic construction of the moduli space

Over the complex numbers, we can construct the moduli space of principally polarised abelian varieties analytically as a quotient

$$\mathrm{Sp}_{2g}(\mathbb{Z})\backslash\mathcal{H}_g.$$

Here $\mathcal{H}_g$ is the **Siegel upper half-space**

$$\mathcal{H}_g = \{Z \in \mathrm{M}_{g\times g}(\mathbb{C}) \mid Z \text{ is symmetric and } \mathrm{Im}\, Z \text{ is positive definite}\}$$

and $\mathrm{Sp}_{2g}(\mathbb{Z})$ is the symplectic group

$$\mathrm{Sp}_{2g}(\mathbb{Z}) = \{M \in \mathrm{GL}_{2g}(\mathbb{Z}) \mid M^t J M = J\} \text{ where } J = \left(\begin{smallmatrix} 0 & 1_g \\ -1_g & 0 \end{smallmatrix}\right).$$

The action of $\mathrm{Sp}_{2g}(\mathbb{Z})$ on $\mathcal{H}_g$ is given by

$$\left(\begin{smallmatrix} A & B \\ C & D \end{smallmatrix}\right) Z = (AZ + B)(CZ + D)^{-1} \text{ where } A, B, C, D \in \mathrm{M}_g(\mathbb{Z}),$$

generalising the standard action of $\mathrm{SL}_2(\mathbb{Z})$ on the upper half-plane by Möbius transformations.

Theorem 5.3 *The quotient $\mathrm{Sp}_{2g}(\mathbb{Z})\backslash\mathcal{H}_g$ is the complex analytic space underlying a quasi-projective variety defined over $\mathbb{Q}$. It is the same as the moduli space of principally polarised abelian varieties over $\mathbb{C}$ defined in the previous section.*

We shall explain how the bijection between $\mathrm{Sp}_{2g}(\mathbb{Z})\backslash\mathcal{H}_g$ and the isomorphism classes of principally polarised abelian varieties over $\mathbb{C}$ is defined. Let (A, λ) be a principally polarised abelian variety of dimension g over $\mathbb{C}$, and let

V/Λ be a complex torus isomorphic to A. The polarisation λ corresponds to a Riemann form ψ on (V, Λ).

Choose a symplectic basis $\{e_1, \ldots, e_g, f_1, \ldots, f_g\}$ for (Λ, ψ), that is, a $\mathbb{Z}$-basis for Λ satisfying

$$\psi(e_j, f_j) = 1 \text{ for all } j, \text{ and}$$

$$\psi(e_j, e_k) = \psi(f_j, f_k) = \psi(e_j, f_k) = 0 \text{ for all } j \neq k.$$

The conditions for ψ to be a Riemann form imply that $e_1, \ldots, e_g$ form a $\mathbb{C}$-basis for V, so we can form a $g \times g$ complex matrix consisting of the coordinates of $f_1, \ldots, f_g$ with respect to $e_1, \ldots, e_g$. This matrix is called a **period matrix** for (A, λ), and the definition of a Riemann form implies that it is in $\mathcal{H}_g$.

Different choices of symplectic basis for Λ may give rise to different period matrices for the same principally polarised abelian variety; this ambiguity is precisely the action of $\mathrm{Sp}_{2g}(\mathbb{Z})$ on $\mathcal{H}_g$.

5.4 The moduli space of elliptic curves

We have already seen that every elliptic curve has a unique principal polarisation, so we can describe $\mathcal{A}_1$ as the moduli space of elliptic curves without mentioning polarisations.

The Siegel upper half-space $\mathcal{H}_1$ is simply the upper half-plane consisting of complex numbers with positive imaginary part, and the symplectic group $\mathrm{Sp}_2(\mathbb{Z})$ is equal to $\mathrm{SL}_2(\mathbb{Z})$. Hence the moduli space of elliptic curves is given analytically by the well-known quotient $\mathrm{SL}_2(\mathbb{Z}) \backslash \mathcal{H}_1$, which is isomorphic to the affine line $\mathbb{C}$.

Over any field, the moduli space of elliptic curves $\mathcal{A}_1$ is isomorphic to the affine line $\mathbb{A}^1$. The point of $\mathbb{A}^1$ associated with an isomorphism class of elliptic curves is given by the j-invariant. For an elliptic curve E in Weierstrass form $y^2 = x^3 + ax + b$, this is given by

$$j(E) = \frac{1728 \cdot 4a^3}{4a^3 + 27b^2}.$$

5.5 Level structures

The moduli spaces $\mathcal{A}_g$ have several deficiencies:

(i) There is no universal principally polarized family of abelian varieties, that is, a family over $\mathcal{A}_g$ such that the map φ defined in property (a) of the moduli space is the identity on $\mathcal{A}_g$.

(ii) The map $\mathcal{H}_g \rightarrow \mathcal{A}_g(\mathbb{C})$ is ramified.

(iii) For $g \geq 2$, $\mathcal{A}_g$ is not smooth.

All of these deficiencies are related to the fact that $\mathrm{Sp}_{2g}(\mathbb{Z})$ contains torsion elements, and to the fact that a principally polarised abelian variety can have non-trivial automorphisms. To avoid these problems, we introduce level structures.

Let N be a positive integer. Recall that the N-torsion of an abelian variety A of dimension g is isomorphic as a group to $(\mathbb{Z}/N\mathbb{Z})^{2g}$. A principal polarisation $\lambda \colon A \rightarrow A^\vee$ induces a symplectic pairing

$$e_N \colon A[N] \times A[N] \rightarrow \mu_N$$

where μ_N is the multiplicative group of roots of unity.

Fix a group isomorphism $\iota \colon \mu_N \rightarrow \mathbb{Z}/N\mathbb{Z}$. A **level-$N$ structure** on (A, λ) is a symplectic basis for $A[N]$ as a $\mathbb{Z}/N\mathbb{Z}$-module with respect to the symplectic pairing $\iota \circ e_N$.

For any N, we can define the **moduli space of principally polarised abelian varieties with level-N structure** in a similar way to the definition of the moduli space of principally polarised abelian varieties. This is an algebraic variety defined over any field which contains μ_N. It is denoted $\mathcal{A}_g(N)$ or $\mathcal{A}_{g,1,N}$ (in the second notation, the 1 indicates that we are dealing with polarisations of degree 1).

For $N \geq 4$, there is a universal principally polarised family of abelian varieties of dimension g with level-N structure on the base $\mathcal{A}_g(N)$ – in other words, a family such that the map φ defined in property (a) is the identity map on $\mathcal{A}_g(N)$. Furthermore $\mathcal{H}_g \rightarrow \mathcal{A}_g(N)$ is the topological universal cover and $\mathcal{A}_g(N)$ is smooth. All of these properties are related to the fact that a principally polarised abelian variety with level-N structure has no non-trivial automorphisms preserving the level structure when $N \geq 4$.

Complex analytically, $\mathcal{A}_g(N)$ is given by $\Gamma(N) \backslash \mathcal{H}_g$ where $\Gamma(N)$ is the group of matrices in $\mathrm{Sp}_{2g}(\mathbb{Z})$ which are congruent to the identity modulo N.

Whenever M divides N, there is a morphism $\mathcal{A}_g(N) \rightarrow \mathcal{A}_g(M)$ which sends a principally polarised abelian variety with level-N structure $(A, \lambda, \{e_1, \ldots, e_{2g}\})$ to the same principally polarised abelian variety with the level-M structure

$$\{(N/M)e_1, \ldots, (N/M)e_{2g}\}.$$

5.6 The moduli space as a Shimura variety

We will not define any of the concepts related to Shimura varieties, but to aid the reader we will explain how the objects defined in this section relate to the definitions in Daw's paper.

The Siegel upper half-space $\mathcal{H}_g$ is an example of a Hermitian symmetric domain. Its group of holomorphic automorphisms is $\mathrm{Sp}_{2g}(\mathbb{R})/\{\pm 1\}$, which is the neutral component of the group of real points of the adjoint algebraic group PGSp_{2g}.

When setting up a Shimura datum, it is convenient to work with GSp_{2g} (the group of matrices which preserve a symplectic form up to a multiplication by a scalar) instead of PGSp_{2g} (the quotient of GSp_{2g} by its centre). This is because the standard $2g$-dimensional representation of GSp_{2g} does not factor through PGSp_{2g}.

There is an equivariant bijection between a certain $\mathrm{GSp}_{2g}(\mathbb{R})$-conjugacy class of morphisms $\mathbb{S} \to \mathrm{GSp}_{2g}$ and the union of the Siegel upper and lower half-spaces $\mathcal{H}_g^{\pm}$; this can be made explicit by means of period matrices. Hence $(\mathrm{GSp}_{2g}, \mathcal{H}_g^{\pm})$ is an example of a Shimura datum.

The groups $\mathrm{Sp}_{2g}(\mathbb{Z})$ and $\Gamma(N)$ are congruence subgroups of $\mathrm{GSp}_{2g}(\mathbb{Q})$ and so the quotients

$$\mathrm{Sp}_{2g}(\mathbb{Z}) \backslash \mathcal{H}_g = \mathcal{A}_g(\mathbb{C}), \quad \Gamma(N) \backslash \mathcal{H}_g = \mathcal{A}_g(N)(\mathbb{C})$$

are connected components of Shimura varieties. In fact $\mathcal{A}_g$ is itself a Shimura variety with a single connected component but $\mathcal{A}_g(N)$ is a component of a Shimura variety which may have multiple connected components.

The fact that $\mathrm{Sp}_{2g}(\mathbb{Z}) \backslash \mathcal{H}_g$ is the complex analytic space of an algebraic variety defined over $\mathbb{Q}$ (Theorem 5.3) is an example of a general theorem on Shimura varieties, the existence of canonical models.

5.7 Definability of theta functions

We have mentioned that

$$\mathcal{A}_g(\mathbb{C}) \cong \mathrm{Sp}_{2g}(\mathbb{Z}) \backslash \mathcal{H}_g$$

is a quasi-projective variety. We can realise it as a subset of projective space by using theta functions, which are holomorphic functions $\mathcal{H}_g \to \mathbb{C}$ satisfying suitable transformation rules with respect to the action of $\mathrm{Sp}_{2g}(\mathbb{Z})$. (When $g = 1$, the classical j-function is an example of such a function.)

There is a standard fundamental set in $\mathcal{H}_g$ for the action of $\mathrm{Sp}_{2g}(\mathbb{Z})$, called the **Siegel fundamental set** and denoted $\mathcal{F}_g$. This is a semialgebraic set and by a theorem of Peterzil and Starchenko [PS13], the restrictions of theta functions to $\mathcal{F}_g$ are definable in the o-minimal structure $\mathbb{R}_{an,exp}$ generated by restricted analytic functions and the real exponential function. This is the first ingredient required to use o-minimality to study subvarieties of $\mathcal{A}_g$.

6 Complex multiplication

An abelian variety is said to have complex multiplication if its endomorphism algebra is as large as possible. Abelian varieties with complex multiplication are important because they have special arithmetic properties. These arithmetic properties are used in the definition of canonical models of Shimura varieties. From the point of view of the André–Oort conjecture, the special points on the moduli space of principally polarised abelian varieties are precisely the points which correspond to abelian varieties with complex multiplication.

6.1 Definition of complex multiplication

Let A be an abelian variety over an algebraically closed field of characteristic zero. An **endomorphism** of A means a morphism $A \to A$, and the endomorphisms of A form a ring $\operatorname{End} A$.

Suppose that A is simple. There is a classification of the possible endomorphism rings of A, due to Albert. In particular $\operatorname{End} A$ has rank at most $2 \dim A$ as a $\mathbb{Z}$-module.

A simple abelian variety A is said to **have complex multiplication** (or **CM**) if $\operatorname{End} A$ is commutative and has rank equal to $2 \dim A$. (A non-simple abelian variety is said to have CM if it is isogenous to a product of simple abelian varieties with CM.)

If A is a simple abelian variety with complex multiplication, then $\operatorname{End} A \otimes_{\mathbb{Z}} \mathbb{Q}$ is a **CM field**. This means a totally complex number field F which contains a totally real subfield F_0 such that $[F : F_0] = 2$; equivalently it is a number field F such that the automorphism of F induced by complex conjugation is non-trivial and independent of the choice of embedding $F \hookrightarrow \mathbb{C}$.

6.2 Construction of CM abelian varieties

Let F be a number field. An **order** in F is a subring R of F which is finitely generated as a $\mathbb{Z}$-module and such that $\mathbb{Q}R = F$. For example, if $F = \mathbb{Q}(i)$, then the ring of integers $\mathbb{Z}[i]$ is an order but there are also smaller orders $\mathbb{Z}[Ni]$ for any positive integer N.

If F is a CM field of degree $2g$, we define a **CM type** Φ for F to be a set of g embeddings $F \hookrightarrow \mathbb{C}$ containing one embedding from each conjugate pair. Given a CM type Φ, an order $R \subset F$ and an ideal $I \subset R$, we can define a lattice $\Lambda \subset \mathbb{C}^g$ as the image of I under the map

$$(\phi_1, \ldots, \phi_g) \colon F \to \mathbb{C}^g$$

where the CM type Φ is $\{\phi_1, \ldots, \phi_g\}$. Then $\mathbb{C}^g / \Lambda$ is an abelian variety with complex multiplication, whose endomorphism ring is equal to R.

Conversely, every abelian variety with complex multiplication has the above form for some order R, CM type Φ and ideal I.

It follows that if we fix the order R, there are finitely many isomorphism classes of abelian varieties with complex multiplication whose endomorphism ring is isomorphic to R, because there are finitely many CM types for the field $R \otimes_{\mathbb{Z}} \mathbb{Q}$ and the ideal class group of R is finite. Since there are countably many CM fields and each contains countably many orders this implies that there are countably many isomorphism classes of abelian varieties with complex multiplication.

6.3 Arithmetic properties of CM abelian varieties

Every abelian variety with complex multiplication is defined over a number field. Note that this field of definition is not equal to the CM field F.

The arithmetic of the CM field F is closely related to the arithmetic of abelian varieties with complex multiplication by F. In particular, the action of $\mathrm{Gal}(\bar{\mathbb{Q}}/\mathbb{Q})$ on the isomorphism classes of abelian varieties with complex multiplication by F, and on their torsion points, can be described in terms of the class field theory of F.

6.4 Complex multiplication and transcendence

Because an abelian variety A with complex multiplication is isomorphic to $\mathbb{C}^g / \Lambda$ where Λ, as described above, consists of points whose coordinates are algebraic numbers, any period matrix for A has entries in $\bar{\mathbb{Q}}$. In fact the property that both A and its period matrices are defined over $\bar{\mathbb{Q}}$ characterises abelian varieties with complex multiplication; this generalises Schneider's theorem on the transcendence of the j-invariant.

Theorem 6.1 ([Shi92]) *Let A be a complex abelian variety and Z a period matrix for A.*

A has complex multiplication if and only if A and Z are both defined over $\bar{\mathbb{Q}}$.

7 The Ax–Lindemann–Weierstrass theorem for abelian varieties

In the final part of these notes, we discuss the Ax–Lindemann–Weierstrass theorem for abelian varieties, and give a proof of this theorem using o-minimal geometry and combining ideas of Pila, Ullmo and Yafaev. We recall the statement of this theorem.

Theorem 7.1 *Let*

- *A be an abelian variety of dimension g over $\mathbb{C}$,*
- *$\pi : \mathbb{C}^g \to A$ be the exponential map,*
- *V be a complex algebraic subvariety of A, and*
- *Y be a maximal irreducible complex algebraic subvariety contained in $\pi^{-1}(V)$.*

Then $\pi(Y)$ is a translate of an abelian subvariety of A.

In the statement of the theorem, when we say that Y is a **maximal** irreducible algebraic subvariety in $\pi^{-1}(V)$, we mean that there is no irreducible algebraic subvariety Y' of $\mathbb{C}^g$ such that $Y \subset Y' \subset \pi^{-1}(V)$ and $Y \neq Y'$.

The same theorem holds with "abelian variety" replaced by "torus" (meaning $(\mathbb{C}^\times)^n$) or indeed "commutative algebraic group over $\mathbb{C}$", and $\pi : \mathbb{C}^g \to A$ by the appropriate exponential map. These results are implied by (and are weaker than) the so-called Ax–Schanuel theorem and its generalisation to commutative algebraic groups, proved by differential-algebraic methods in [Ax71] and [Ax72]. For example, Theorem 7.1 is equivalent to the following transcendence statement, which can be deduced from Theorem 3 of [Ax72].

Theorem 7.2 *Let A be an abelian variety of dimension g over $\mathbb{C}$ and let $\pi : \mathbb{C}^g \to A$ be the exponential map. Let $f_1, \ldots, f_n : A \to \mathbb{C} \cup \{\infty\}$ be meromorphic functions on A such that $[f_1 : \cdots : f_n]$ defines a projective embedding of A.*

Let Y be an irreducible complex algebraic variety and let $y_1, \ldots, y_g \in \mathbb{C}[Y]$ be regular functions on Y. Suppose that the image of $\pi \circ (y_1, \ldots, y_g) : Y \to A$ is not contained in any translate of a proper abelian subvariety of A.

Then the set of meromorphic functions

$$\{f_j \circ \pi \circ (y_1, \ldots, y_g) : Y \to \mathbb{C} \cup \{\infty\} \mid j = 1, \ldots, n\}$$

has transcendence degree g over $\mathbb{C}$.

Theorem 7.1 is an ingredient in the proof of the Pila–Zannier proof of the Manin–Mumford conjecture. The Pila–Zannier approach to the André–Oort conjecture requires an analogous theorem for Shimura varieties. Such a theorem was proved for products of modular curves by Pila [Pil11], for compact Shimura varieties by Ullmo and Yafaev [UY13], for the moduli space of principally polarised abelian varieties by Pila and Tsimerman [PT12], and finally for general Shimura varieties by Klingler, Ullmo and Yafaev [KUY13].

7.1 Outline of proof

Let $\Lambda = \ker \pi$, which is a lattice in $\mathbb{C}^g$, and let $\mathcal{F} \subset \mathbb{C}^g$ be the interior of a fundamental parallelepiped for Λ. We will apply the Pila–Wilkie theorem to the set

$$\Sigma = \{x \in \mathbb{C}^g \mid (Y + x) \cap \mathcal{F} \neq \emptyset \text{ and } Y + x \subset \pi^{-1}(V)\}.$$

Note that in most papers on this subject, Σ is defined as

$$\{x \in \mathbb{C}^g \mid \dim((Y + x) \cap \mathcal{F} \cap \pi^{-1}(V)) = \dim Y\},$$

which is the same set as the previous definition.

Of the two conditions in the definition of Σ, the important one for proving Theorem 7.1 is the second condition,

$$Y + x \subset \pi^{-1}(V).$$

If W is an irreducible semialgebraic set of points satisfying this condition and containing 0, then the maximality of Y implies that $Y + W = Y$. So if there exists such a W with positive dimension then we can make the key conclusion that the stabiliser of Y has positive dimension.

However the set

$$\{x \in \mathbb{C}^g \mid Y + x \subset \pi^{-1}(V)\}$$

is usually not definable in an o-minimal structure because it has infinitely many connected components. Adding the condition $(Y+x) \cap \mathcal{F} \neq \emptyset$ ensures that Σ is definable in the o-minimal structure $\mathbb{R}_{an}$, and so we can apply the Pila–Wilkie theorem to Σ.

The proof of Theorem 7.1 has the following steps:

(1) Show that the number of points of $\Sigma \cap \Lambda$ of height up to T grows at least linearly with T.
(2) Apply the Pila–Wilkie theorem to deduce that Σ contains a semialgebraic set of positive dimension.

(3) Deduce that the stabiliser of Y has positive dimension, and show that the image of this stabiliser under $\pi : \mathbb{C}^g \to A$ is equal to the stabiliser of the Zariski closure of $\pi(Y)$ (this step uses the maximality of Y).

(4) Quotient out by the stabiliser of Y and complete the proof of Theorem 7.1 by applying the argument again to the quotient.

The proofs of the Ax–Lindemann–Weierstrass theorem for Shimura varieties all follow the same structure as above. Step 1 is much harder for Shimura varieties than for abelian varieties, steps 2 and 3 are essentially the same and step 4 must be replaced by a more complicated argument using monodromy.

8 Relationship between semialgebraic and complex algebraic sets

Before we prove Theorem 7.1, we first need a lemma saying that a maximal complex algebraic variety contained in a complex analytic set Z is in fact maximal among connected irreducible semialgebraic subsets of Z.

In the course of the proof of Theorem 7.1, we will construct a semialgebraic set Y' such that $Y \subset Y' \subset \pi^{-1}(V)$. We wish to use the maximality of Y to deduce that $Y = Y'$, but our hypothesis says that Y is maximal among complex algebraic subvarieties contained in $\pi^{-1}(V)$ and Y' is only semialgebraic. The following lemma deals with this problem.

The lemma as we give it below is Lemma 4.4.1 of [PT13] which seems to be the most elegant statement of this type; a weaker version, sufficient for the applications to André–Oort, was Lemma 4.2.1 in [PZ08].

Lemma 8.1 *Let Z be a complex analytic set in $\mathbb{C}^g$ and $X \subset Z$ a connected irreducible real semialgebraic subset. Then there is a complex algebraic variety X' such that $X \subset X' \subset Z$.*

In the lemma, we identify $\mathbb{C}^g$ with $\mathbb{R}^{2g}$ using the real and imaginary parts of the coordinates. We say that a semialgebraic set $X \subset \mathbb{R}^{2g}$ is **irreducible** if it cannot be written as a union $X = X_1 \cup X_2$ of two proper subsets which are closed in the topology induced on X by the Zariski topology of real algebraic sets in $\mathbb{R}^{2g}$.

Proof Let S be the real Zariski closure of X in $\mathbb{R}^{2g}$. Any cell of maximum dimension in X is Zariski dense in S. This implies that S is geometrically irreducible (that is, S is irreducible over as an algebraic set over $\mathbb{C}$) and that $\dim S = \dim X$.

Define $f\colon \mathbb{C}^{2g} \to \mathbb{C}^g$ by

$$f(x_1, y_1, \ldots, x_g, y_g) = (x_1 + iy_1, \ldots, x_g + iy_g)$$

and let ι be the inclusion $\mathbb{R}^{2g} \to \mathbb{C}^{2g}$. Then the composite $f \circ \iota$ is the isomorphism we are using to identify $\mathbb{R}^{2g}$ with $\mathbb{C}^g$.

Let S_1 be the closure of $\iota(S)$ in the complex Zariski topology on $\mathbb{C}^{2g}$ (S_1 is the "extension of scalars" of S from $\mathbb{R}$ to $\mathbb{C}$). Let $S_2 = f(S_1)$.

We have the following diagram:

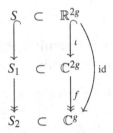

By Chevalley's theorem, S_2 is constructible in the complex Zariski topology on $\mathbb{C}^g$. Hence the closure of S_2 in the Euclidean topology is a complex algebraic set. We shall denote this closure of S_2 by X', and show that it has the required property $X \subset X' \subset Z$.

It is easy to show that $X \subset X'$: we have

$$X \subset S \subset S_2 \subset X'.$$

To show that $X' \subset Z$, we shall apply the following claim to $f^{-1}(Z) \cap S_1$. This contains $\iota(X)$ by the hypothesis $X \subset Z$ and by the definition of S_1. Hence by the claim, $f^{-1}(Z) \cap S_1 = S_1$ and so $f(S_1) \subset Z$. Since Z is closed and $f(S_1)$ is dense in X' for the Euclidean topology, we can conclude that $X' \subset Z$.

Claim   There are no proper closed analytic subsets of S_1 containing $\iota(X)$.

To prove the claim, let W be the smallest analytic subset of S_1 containing $\iota(X)$.

Let x be a point in a cell of maximum dimension in X such that W is smooth at $x_1 = \iota(x)$. To see that such a point exists, observe that the smooth points of W form a non-empty open subset of W. By the minimality of W, $\iota(X) \not\subset W^{\mathrm{sing}}$ so the set of points $x \in X$ such that $\iota(x)$ is a smooth point of W is a non-empty open subset of X. In particular, this set intersects at least one cell of maximum dimension in X.

Since x is in a maximum dimensional cell of X, locally near x, S and X coincide. Since $\iota(X) \subset W$, we deduce that

$$T_{x_1}\iota(S) \subset T_{x_1}W.$$

Here $\iota(S)$ is a real algebraic set and W is complex analytic, so $T_{x_1}\iota(S)$ is a real vector space and $T_{x_1}W$ is a complex vector space. Thus $T_{x_1}W$ contains the complex vector space $\mathbb{C}T_{x_1}\iota(S)$ generated by $T_{x_1}\iota(S)$.

By the definition of S_1,

$$\mathbb{C}T_{x_1}\iota(S) = T_{x_1}S_1$$

so we have

$$T_{x_1}S_1 \subset T_{x_1}W.$$

Since W is smooth at x_1, $\dim T_{x_1}W = \dim W$. So we get

$$\dim T_{x_1}S_1 \leq \dim T_{x_1}W = \dim_{x_1}W \leq \dim S_1 \leq \dim T_{x_1}S_1.$$

These inequalities must be equalities so

$$\dim_{x_1}W = \dim S_1.$$

Combined with the fact that $W \subset S_1$ and that S_1 is irreducible, we deduce that $W = S_1$. $\qquad\square$

9 Proof of the Ax–Lindemann–Weierstrass theorem for abelian varieties

In this section we prove Theorem 7.1. We will follow the outline of the proof and use the notation from section 7.

We suppose that $\dim Y > 0$ – otherwise the theorem is trivial because any point in A is translate of the abelian subvariety $\{0\}$. We also assume that V is the Zariski closure in A of $\pi(Y)$ – replacing a larger subvariety V by the Zariski closure of $\pi(Y)$ will not change the fact that Y is a maximal complex algebraic subvariety in $\pi^{-1}(V)$.

9.1 Lattice points in Σ

Recall that Λ denotes the kernel of $\pi \colon \mathbb{C}^g \to A$, which is a lattice in $\mathbb{C}^g$. Fix a $\mathbb{Z}$-basis $e_1, \ldots, e_{2g}$ for Λ and use this to define the height of elements of Λ:

$$H(a_1 e_1 + \cdots + a_{2g}e_{2g}) = \max(|a_1|, \ldots, |a_{2g}|).$$

Observe that all points $x \in \Lambda$ satisfy $Y + x \subset \pi^{-1}(V)$ because $\pi^{-1}(V)$ is Λ-invariant. Hence

$$\Sigma \cap \Lambda = \{x \in \Lambda \mid (Y + x) \cap \mathcal{F} \neq \emptyset\}$$

and in order to count lattice points in Σ, it suffices to count points in the latter set.

Lemma 9.1   *There exists $T_0 \in \mathbb{R}$ such that for all $T > T_0$,*

$$\#\{x \in \Sigma \cap \Lambda \mid H(x) \leq T\} \geq T/2.$$

Proof   Since Y is an irreducible affine algebraic variety, it is path-connected and unbounded with respect to the usual norm on $\mathbb{C}^g$. Hence we can find a continuous function $\gamma : [0, \infty) \to Y$ whose image is unbounded.

Each time the image of γ crosses a boundary from one fundamental domain $\mathcal{F} - x$ to another $\mathcal{F} - x'$ (with $x, x' \in \Lambda$), the heights of x and x' differ by at most 1. So the heights of points in

$$\Lambda_\gamma = \{x \in \Lambda \mid \mathrm{im}\, \gamma \cap (\mathcal{F} - x) \neq \emptyset\}$$

form a set of consecutive integers.

Since γ is unbounded, Λ_γ contains points of arbitrarily large height. Hence there is some h_0 such that for every integer $h > h_0$, Λ_γ contains at least one point of height h.

By the observation above the lemma, $\Lambda_\gamma \subset \Sigma \cap \Lambda$. Hence the lemma is proved with $T_0 = 2h_0$. $\square$

Lemma 9.1 is straightforward to prove, but the analogous statement for Shimura varieties (proved by Klingler, Ullmo and Yafaev) is much more difficult. This is because the arithmetic group acting on the uniformising space of a Shimura variety (the analogue of the lattice Λ) is non-commutative and the heights of elements of this group grow exponentially instead of linearly with respect to the word metric. Hence in order to prove an analogue of Lemma 9.1 it is necessary to show that Y intersects exponentially many fundamental domains (instead of just one) at a given distance from the base point.

9.2 Applying the Pila–Wilkie theorem

We will now apply the Pila–Wilkie theorem to deduce that Σ contains a real semialgebraic set of positive dimension.

In order to state the Pila–Wilkie theorem, we make the following definition: if X is a subset of $\mathbb{R}^n$, let X^{alg} denote the union of all connected positive-dimensional real semialgebraic subsets contained in X.

Theorem 9.2 ([PW06] Theorem 1.8) *Let $X \subset \mathbb{R}^n$ be a set definable in $\mathbb{R}_{\mathrm{an}}$ (or any o-minimal structure). Let $\epsilon > 0$.*
There exists a constant c, depending only on X and ϵ, such that for all $T \geq 1$,

$$\#\{x \in X - X^{\mathrm{alg}} \mid x \in \mathbb{Q}^n \text{ and } H(x) \leq T\} \leq cT^{\epsilon}.$$

In applying this theorem, we will identify $\mathbb{C}^g$ with $\mathbb{R}^{2g}$ by identifying the basis $e_1, \ldots, e_{2g}$ which we chose for Λ with the standard basis for $\mathbb{R}^{2g}$. Thus Λ is identified with $\mathbb{Z}^{2g}$ and the height of points in Λ which we defined above is identified with the naive height on $\mathbb{Z}^{2g}$.

In order to apply Theorem 9.2 to Σ, we need to check that Σ is definable in $\mathbb{R}_{\mathrm{an}}$, the o-minimal structure generated by restricted analytic functions.

Lemma 9.3 *The set Σ is definable in $\mathbb{R}_{\mathrm{an}}$.*

Proof Note that Σ is equal to

$$\{x \in \mathbb{C}^g \mid (Y + x) \cap \mathcal{F} \neq \emptyset \text{ and } (Y + x) \cap \mathcal{F} \subset \pi^{-1}(V) \cap \mathcal{F}\}.$$

This holds because $Y + x$ and $\pi^{-1}(V)$ are both complex analytic sets, with $Y + x$ irreducible, and $(Y + x) \cap \mathcal{F}$ is an open subset of $Y + x$. Hence if $(Y + x) \cap \mathcal{F}$ is non-empty and is contained in $\pi^{-1}(V)$, then by analytic continuation $Y + x$ is contained in $\pi^{-1}(V)$.

The set above is definable in $\mathbb{R}_{\mathrm{an}}$ because $(Y + x) \cap \mathcal{F}$ and $\pi^{-1}(V) \cap \mathcal{F}$ are each defined by analytic functions on $\mathbb{C}^g$ restricted to the parallelepiped $\mathcal{F}$. □

Applying Theorem 9.2 together with the lower bound from Theorem 9.1 shows that $\Sigma^{\mathrm{alg}} \cap \Lambda$ is non-empty. Let W be a connected irreducible positive-dimensional semialgebraic set, such that $W \subset \Sigma$ and W contains some point $w_0 \in \Lambda$.

9.3 The stabilisers of Y and V

Let $\Theta \subset \mathbb{C}^g$ denote the stabiliser of Y (which is a vector subspace of $\mathbb{C}^g$), and let $B \subset A$ denote the identity component of the stabiliser of V (which is an abelian subvariety of A). We will show that $W - w_0 \subset \Theta$ and that $\pi(\Theta) = B$. Hence both Θ and B have positive dimension.

Lemma 9.4 *If $W \subset \Sigma$ is a connected irreducible semialgebraic set and $w_0 \in W \cap \Lambda$, then $Y + W - w_0 = Y$.*

Proof By the definition of Σ, every point $w \in W$ satisfies $Y + w \subset \pi^{-1}(V)$. Hence

$$Y + W \subset \pi^{-1}(V).$$

Since $\pi^{-1}(V)$ is Λ-invariant, we also have

$$Y + W - w_0 \subset \pi^{-1}(V).$$

But $Y + W - w_0$ is a connected irreducible real semialgebraic set so Lemma 8.1 implies that there is some irreducible complex algebraic variety Y' such that

$$Y \subset Y + W - w_0 \subset Y' \subset \pi^{-1}(V).$$

By hypothesis, Y is a maximal algebraic variety contained in $\pi^{-1}(V)$. Hence

$$Y = Y + W - w_0 = Y'. \qquad \qquad \square$$

Lemma 9.5 $\pi(\Theta) = B$.

Proof First we show that $\pi(\Theta) \subset B$, using the minimality of V. Suppose that $x \in \Theta$. Then

$$Y + x = Y \subset \pi^{-1}(V)$$

so $Y \subset \pi^{-1}(V) - x$.
    Hence

$$\pi(Y) \subset V \cap (V - \pi(x)).$$

The set $V \cap (V - \pi(x))$ is an algebraic subvariety of A, so the assumption that V is the Zariski closure of $\pi(Y)$ implies that $V = V - \pi(x)$.
    Thus $\pi(\Theta)$ stabilises V. Since Θ is connected in the Euclidean topology, $\pi(\Theta)$ is connected in the Euclidean and hence also in the Zariski topology, so $\pi(\Theta) \subset B$.
    Now we show that $B \subset \pi(\Theta)$. Let Θ' be the identity component of $\pi^{-1}(B)$ in the Euclidean topology, and note that the image $\pi(\Theta)$ is an analytic subgroup of B of the same dimension as B, so must be equal to B. Hence it will suffice to show that $\Theta' \subset \Theta$.
    Since Θ' stabilises $\pi^{-1}(V)$, we have

$$Y + \Theta' \subset \pi^{-1}(V).$$

But $Y + \Theta'$ is an irreducible complex algebraic set containing Y, so the maximality of Y implies that $Y + \Theta' = Y$ as required. $\qquad\square$

9.4 Conclusion of proof

We have proved that $\dim B > 0$ where B is the stabiliser of V, which is an abelian subvariety of A.

Let A' denote the abelian subvariety A/B. We have proved that $B = \pi(\Theta)$ so we have a diagram of quotient maps

$$
\begin{array}{ccc}
\mathbb{C}^g & \xrightarrow{\ \tilde{q}\ } & \mathbb{C}^g/\Theta \\
\ \downarrow{\scriptstyle \pi} & & \ \downarrow{\scriptstyle \pi'} \\
A & \xrightarrow{\ q\ } & A'
\end{array}
$$

Let $V' = q(V)$ and let Y' be the closure of $\tilde{q}(Y)$ in $\mathbb{C}^g/\Theta$. Note that $V = q^{-1}(V')$ and $Y = \tilde{q}^{-1}(Y')$.

Now Y' is a maximal irreducible algebraic subvariety in $\pi'^{-1}(V')$. To prove this, suppose that there were some irreducible algebraic subvariety Z' of $\mathbb{C}^g/\Theta$ such that $Y' \subset Z' \subset \pi'^{-1}(V')$. Then

$$
Y \subset \tilde{q}^{-1}(Z') \subset \pi^{-1}(V)
$$

and each irreducible component of $\tilde{q}^{-1}(Z')$ is an algebraic subvariety of $\mathbb{C}^g$. So the maximality of Y implies that irreducible component of $\tilde{q}^{-1}(Z')$ containing Y is equal to Y, and hence $Z' = Y'$.

If $\dim Y' > 0$ then we can apply the above argument to (A', V', Y') to deduce that the stabiliser of V' in A' has positive dimension. But the preimage of this stabiliser under q stabilises V, so this contradicts the fact that $\ker q$ is equal to the stabiliser of V.

Hence Y' is a point and $\pi(Y)$ is a translate of the abelian subvariety $B \subset A$ as required.

References

[Ax71]   J. Ax. On Schanuel's conjectures. *Ann. of Math. (2)*, 93:252–268, 1971.

[Ax72]   J. Ax. Some topics in differential algebraic geometry. I. Analytic subgroups of algebraic groups. *Amer. J. Math.*, 94:1195–1204, 1972.

[BL92]   Christina Birkenhake and Herbert Lange. *Complex abelian varieties*, volume 302 of *Grundlehren der Mathematischen Wissenschaften [Fundamental Principles of Mathematical Sciences]*. Springer-Verlag, Berlin, 1992.

[KUY13] B. Klingler, E. Ullmo, and A. Yafaev. The hyperbolic Ax–Lindemann–Weierstrass conjecture. Preprint, available at arxiv.org/abs/1307.3965, 2013.

[Mas84] D. W. Masser. Small values of the quadratic part of the Néron-Tate height on an abelian variety. *Compositio Math.*, 53(2):153–170, 1984.

[Mil86] J. S. Milne. Abelian varieties. In *Arithmetic geometry (Storrs, Conn., 1984)*, pages 103–150. Springer, New York, 1986.

[Mil06] J. S. Milne. Complex multiplication. Available at http://www.jmilne.org/math/CourseNotes/cm.html, 2006.

[Mil08] J. S. Milne. Abelian varieties (v2.00). Available at http://www.jmilne.org/math/CourseNotes/av.html, 2008.

[Mum65] D. Mumford. *Geometric invariant theory*. Ergebnisse der Mathematik und ihrer Grenzgebiete, Neue Folge, Band 34. Springer-Verlag, Berlin, 1965.

[Mum70] David Mumford. *Abelian varieties*. Tata Institute of Fundamental Research Studies in Mathematics, No. 5. Published for the Tata Institute of Fundamental Research, Bombay, 1970.

[Pil11] J. Pila. O-minimality and the André–Oort conjecture for $\mathbb{C}^n$. *Ann. of Math. (2)*, 173(3):1779–1840, 2011.

[PS13] Y. Peterzil and S. Starchenko. Definability of restricted theta functions and families of abelian varieties. *Duke Math. J.*, 162(4):731–765, 2013.

[PT12] J. Pila and J. Tsimerman. Ax–Lindemann for $\mathcal{A}_g$. Preprint, available at http://arxiv.org/abs/1206.2663, 2012.

[PT13] J. Pila and J. Tsimerman. The André–Oort conjecture for the moduli space of abelian surfaces. *Compositio Math.*, 149(2):204–216, 2013.

[PW06] J. Pila and A. J. Wilkie. The rational points of a definable set. *Duke Math. J.*, 133(3):591–616, 2006.

[PZ08] J. Pila and U. Zannier. Rational points in periodic analytic sets and the Manin–Mumford conjecture. *Rend. Lincei (9) Mat. Appl.*, 19(2):149–162, 2008.

[Shi92] Hironori Shiga. On the transcendency of the values of the modular function at algebraic points. *Astérisque*, (209):16, 293–305, 1992. Journées Arithmétiques, 1991 (Geneva).

[Sil94] Joseph H. Silverman. *Advanced topics in the arithmetic of elliptic curves*, volume 151 of *Graduate Texts in Mathematics*. Springer-Verlag, New York, 1994.

[UY13] E. Ullmo and A. Yafaev. Hyperbolic Ax–Lindemann theorem in the cocompact case. *Duke Math. J. (to appear)*, available at http://arxiv.org/abs/1209.0939, 2013.

5

The André-Oort conjecture via o-minimality

Christopher Daw*

1 Introduction

Shimura varieties are a distinguished class of algebraic varieties that parameterise important objects from linear algebra called *Hodge structures*. Often these Hodge structures correspond to families of so-called *Abelian varieties*.

Additional structure on a Shimura variety S arises through the existence of certain *algebraic correspondences* on S, i.e. subvarieties of $S \times S$, called *Hecke correspondences*. We can think of these as one-to-many maps

$$T : S \to S.$$

We endow S with a set of so-called *special subvarieties*, defined as the set of all connected components of *Shimura subvarieties* and the irreducible components of their images under Hecke correspondences. This is analogous to the case of Abelian varieties (respectively *algebraic tori*), where special subvarieties are the translates of *Abelian subvarieties* (respectively *subtori*) by *torsion points*. A key property of special subvarieties is that connected components of their intersections are themselves special subvarieties. Thus, any subvariety Y of S is contained in a smallest special subvariety. If this happens to be a connected component of S itself, then we say that Y is *Hodge generic* in S.

We refer to the special subvarieties of dimension zero as *special points*. Special subvarieties contain a Zariski (in fact, analytically) dense set of special points. The *André-Oort conjecture* predicts that this property characterises special subvarieties:

* University College London, Department of Mathematics, Gower Street, WC1E 6BT London, United Kingdom, e-mail: c.daw@ucl.ac.uk

O-Minimality and Diophantine Geometry, ed. G. O. Jones and A. J. Wilkie. Published by Cambridge University Press. © Cambridge University Press 2015.

Conjecture 1.1 (André-Oort)   Let S be a Shimura variety and let Σ be a set of special points contained in S. Every irreducible component of the Zariski closure of $\cup_{s \in \Sigma} s$ in S is a special subvariety.

A connected component of S arises as a quotient $\Gamma \backslash D$, where D is a certain type of complex manifold called a *Hermitian symmetric domain*, and Γ is a certain type of discrete subgroup of $\mathrm{Hol}(D)^+$ called a *congruence subgroup*. From now on, we will use S to denote this component.

By [10], §3, there exists a semi-algebraic *fundamental domain* $\mathcal{F} \subset D$ for the action of Γ. By [10], Theorem 1.2, when the uniformisation map

$$\pi : D \to S$$

is restricted to $\mathcal{F}$, one obtains a function definable in the o-minimal structure $\mathbb{R}_{an,exp}$. Through these observations, the André-Oort conjecture becomes amenable to tools from o-minimality.

The purpose of this article is to explain the so-called *Pila-Zannier strategy* for proving the André-Oort conjecture. This strategy first arose in a proof of the *Manin-Mumford conjecture* [20] and was first adapted to Shimura varieties by Pila [16]. We will follow the outline given by Ullmo [24] for $\mathcal{A}_6^r$, where $\mathcal{A}_g$ is the moduli space for *principally polarised* Abelian varieties of dimension g.

The first step is to show that, if Y is an irreducible, Hodge generic subvariety of S, then the union of all positive-dimensional, special subvarieties contained in Y is not Zariski dense in Y. The second step is to show that all but finitely many special points in Y lie on a positive-dimensional, special subvariety contained in Y.

Both steps require the *hyperbolic Ax-Lindemann-Weierstrass conjecture*, a geometric statement itself amenable to proof via o-minimality. Other articles in these proceedings will explain this conjecture in detail along with its analogue in the case of an Abelian variety. Let us just mention that the conjecture was first proven in the cocompact case by Ullmo and Yafaev [25], then by Pila and Tsimerman for $\mathcal{A}_g$ [18], and finally by Klingler, Ullmo and Yafaev in the general case [10].

Ullmo demonstrates the first step in his article [24]. Therefore, the focus of this article will be the second step. The strategy will be to compare lower bounds for the size of Galois orbits of special points with upper bounds for the height of their pre-images in the fundamental domain. One concludes by applying the *Pila-Wilkie counting theorem* [19], which states that the number of algebraic points of degree at most k and height at most T, in the complement of all connected, positive-dimensional, semi-algebraic subsets of a set X, definable in an o-minimal structure, is $\ll_{\epsilon,k,X} T^\epsilon$.

First, however, we will provide a brief introduction to the theory of Shimura varieties, as formulated by Deligne in his foundational articles [4] and [5]. Our introduction is not by any means intended to be a full treatment of the topic but rather a preparatory guide for graduate students approaching it for the first time. We refer the reader to [12] for a comprehensive account of Shimura varieties and for further details regarding the topics introduced here.

2 Hermitian symmetric domains

We are primarily interested in the connected components of Shimura varieties. These initially arise as quotients $\Gamma \backslash D$, where D is a certain type of complex manifold called a *Hermitian symmetric domain*, and Γ is a *congruence subgroup*, acting via holomorphic automorphisms. The protypical example is the case of the upper half-plane

$$D = \mathbb{H} := \{ z \in \mathbb{C} : \Im(z) > 0 \}$$

and $\Gamma = \mathrm{SL}_2(\mathbb{Z})$, where any element of $\mathrm{SL}_2(\mathbb{R})$ acts on $\mathbb{H}$ by

$$\begin{pmatrix} a & b \\ c & d \end{pmatrix} \cdot z = \frac{az + b}{cz + d}.$$

We refer the reader to [12], §1 for a detailed introduction to Hermitian symmetric domains. We merely summarise the key points. Unfortunately, the definition is not particularly enlightening:

Definition 2.1 A Hermitian symmetric domain is a connected complex manifold D such that

- D is equipped with a Hermitian metric.
- The group $\mathrm{Aut}(D)$ of holomorphic isometries acts transitively on D.
- There exists a point $\tau \in D$ and an involution $\varphi \in \mathrm{Aut}(D)$ such that τ is an isolated fixed point of φ.
- D is of non-compact type.

For any topological group G, we denote its *neutral component* by G^+. By this we mean the connected component of G containing the identity element $\mathrm{id} \in G$. By [12], Lemma 1.5, $\mathrm{Aut}(D)^+$ acts transitively on D and, by [12], Proposition 1.6, it coincides with $\mathrm{Hol}(D)^+$, where $\mathrm{Hol}(D)$ denotes the larger group of all holomorphic automorphisms. Note that, given the transitivity of the $\mathrm{Aut}(D)$ action, the third condition is true for all points $\tau \in D$.

Returning to our earlier example,

$$\mathrm{Hol}(\mathbb{H}) = \mathrm{SL}_2(\mathbb{R})/\{\pm\mathrm{id}\}.$$

Since $\mathrm{SL}_2(\mathbb{R})$ is connected, so is $\mathrm{Hol}(\mathbb{H})$ and it therefore coincides with $\mathrm{Aut}(\mathbb{H})$. The element

$$\varphi := \begin{pmatrix} 0 & 1 \\ -1 & 0 \end{pmatrix} \in \mathrm{SL}_2(\mathbb{R})$$

fixes only $i \in \mathbb{H}$, whereas $\varphi^2 = -\mathrm{id}$. Hence, the image of φ in $\mathrm{Aut}(\mathbb{H})$ is an involution of $\mathbb{H}$ with an isolated fixed point.

However, from the definition follows a key property of Hermitian symmetric domains: by [12], Theorem 1.9, if we denote by $\mathbb{U}(\mathbb{R})$ the *circle group* $\{z \in \mathbb{C} : |z| = 1\}$, then for each point $\tau \in D$ there exists a unique homomorphism

$$u_\tau : \mathbb{U}(\mathbb{R}) \to \mathrm{Hol}(D)^+$$

such that, for all $z \in \mathbb{U}(\mathbb{R})$,

- $u_\tau(z)(\tau) = \tau$.
- $u_\tau(z)$ acts as multiplication by z on the tangent plane of D at τ.

For example, consider the point $i \in \mathbb{H}$ and let

$$h_i : \mathbb{U}(\mathbb{R}) \to \mathrm{SL}_2(\mathbb{R}) : z = a + ib \mapsto \begin{pmatrix} a & b \\ -b & a \end{pmatrix}.$$

Then, for all $z \in \mathbb{U}(\mathbb{R})$, $h_i(z)$ fixes i and

$$\frac{d}{dz}\left(\frac{az+b}{-bz+a}\right)\bigg|_i = \frac{a^2+b^2}{(a-bi)^2} = \frac{z}{\bar{z}}.$$

Therefore, if we define

$$u_i : \mathbb{U}(\mathbb{R}) \to \mathrm{SL}_2(\mathbb{R})/\{\pm\mathrm{id}\} : z \mapsto h_i(\sqrt{z}) \bmod \pm \mathrm{id},$$

which is well-defined since $h_i(-1) = -\mathrm{id}$, then $u_i(z)$ acts on the tangent plane of $\mathbb{H}$ at i as multiplication by z.

Furthermore, note that, if $g \in \mathrm{Hol}(D)^+$ and $\tau \in D$, then the uniqueness of $u_{\tau x}$ implies that it must be the conjugate

$$g u_\tau g^{-1} : z \mapsto g u_\tau(z) g^{-1}.$$

Therefore, since $\mathrm{Hol}(D)^+$ acts transitively on D, if we fix a point $\tau_0 \in D$, we have a $\mathrm{Hol}(D)^+$-equivariant bijection between D and the $\mathrm{Hol}(D)^+$-conjugacy class of u_{τ_0}.

3 Conjugacy classes

By [12], Proposition 1.7, for any Hermitian symmetric domain D, there exists a unique, adjoint, semisimple algebraic group G over $\mathbb{R}$ such that

$$G(\mathbb{R})^+ = \mathrm{Hol}(D)^+.$$

By a *linear algebraic group* G over $\mathbb{R}$, we simply mean a group that can be defined as a subgroup of $\mathrm{GL}_n(\mathbb{R})$ by real polynomials in the matrix coefficients. For example, $\mathbb{U}(\mathbb{R})$ is a linear algebraic group over $\mathbb{R}$ whose elements may be realised as those

$$\begin{pmatrix} a & b \\ c & d \end{pmatrix} \in \mathrm{GL}_2(\mathbb{R})$$

such that $a = d$, $b = -c$ and $a^2 + b^2 = 1$ (in particular, $\mathbb{U}(\mathbb{R})$ is contained in $\mathrm{SL}_2(\mathbb{R})$). However, since $\mathbb{U}(\mathbb{R})$ is defined by polynomials, we can think of $\mathbb{U}(\mathbb{R})$ as the *real points* of what is usually considered the algebraic group, which we denote $\mathbb{U}$. Then, for any $\mathbb{R}$-algebra A, $\mathbb{U}(A)$ is simply the group of solutions in A to the above polynomials.

By a *semisimple* algebraic group we mean a connected (for the Zariski topology), linear algebraic group that is isogenous to a product of almost-simple subgroups. By a *simple* algebraic group we mean a connected, linear algebraic group that is not commutative and has no proper, normal, algebraic subgroups other than the identity. By an *almost-simple* subgroup we mean a subgroup that is a simple algebraic group modulo a finite centre. An *isogeny* between semisimple algebraic groups is a surjective morphism with finite kernel. Two semisimple algebraic groups H_1 and H_2 are called *isogenous* if there exist isogenies

$$H_1 \leftarrow G \rightarrow H_2,$$

for some semisimple algebraic group G. This is an equivalence relation. By *adjoint* we are referring to a group with trivial centre and, for a linear algebraic group G, we write G^{ad} for G modulo its centre.

As shown in [12], §1, every representation

$$\mathbb{U}(\mathbb{R}) \rightarrow \mathrm{GL}_n(\mathbb{R})$$

is algebraic i.e. the image is given by polynomials in the matrix entries and can be written $\mathbb{U} \rightarrow \mathrm{GL}_n$. In particular, for any $\tau \in D$, we may consider the homomorphism

$$u_\tau : \mathbb{U}(\mathbb{R}) \rightarrow G(\mathbb{R})^+$$

as an algebraic morphism $u_\tau : \mathbb{U} \to G$, yielding a morphism

$$u_\tau : \mathbb{U}(A) \to G(A)$$

for any $\mathbb{R}$-algebra A.

The group $\mathbb{U}$ is connected, commutative and consists entirely of *semisimple elements*. By the latter condition we mean that, for any representation

$$\mathbb{U} \to \mathrm{GL}_n,$$

any element in the image of $\mathbb{U}(\mathbb{C})$ can be diagonalised by an element of $\mathrm{GL}_n(\mathbb{C})$. The fact that $\mathbb{U}$ is also commutative implies that the elements in the image of $\mathbb{U}(\mathbb{C})$ can be simultaneously diagonalised by a single element of $\mathrm{GL}_n(\mathbb{C})$. We refer to a linear algebraic group of this sort as a *torus*.

For any representation of $\mathbb{U}$, the eigenvalues are given by homomorphisms $\mathbb{U}_{\mathbb{C}} \to \mathbb{G}_m$ called *characters*, where we write $\mathbb{U}_{\mathbb{C}}$ for $\mathbb{U}$ considered as an algebraic group over $\mathbb{C}$ and $\mathbb{G}_m$ for the algebraic group such that, for any $\mathbb{C}$-algebra A,

$$\mathbb{G}_m(A) = A^\times := \{a \in A : a \text{ is invertible in } A\}.$$

The characters are algebraic since, by definition, they are one-dimensional representations. In this case, each character is of the form $z \mapsto z^n$, where $n \in \mathbb{Z}$.

By [12], Theorem 1.21, the homomorphism u_τ always satisfies the following three properties:

- Only the characters $z \mapsto 1$, $z \mapsto z$ and $z \mapsto z^{-1}$ occur in the representation of $\mathbb{U}(\mathbb{R})$ on the Lie algebra $\mathfrak{g}_{\mathbb{C}}$ of $G_{\mathbb{C}}$.
- Conjugation by $u_\tau(-1)$ is a Cartan involution of G.
- $u_\tau(-1)$ maps to a non-trivial element in every simple factor of G.

The *Lie algebra* of $G_{\mathbb{C}}$ is the tangent plane of $G(\mathbb{C})$ at the identity. One definition is the kernel of the map

$$G(\mathbb{C}[\epsilon]) \to G(\mathbb{C})$$

induced by $\epsilon \mapsto 0$, where $c^2 = 1$. Then $G(\mathbb{C})$ acts on $\mathfrak{g}_{\mathbb{C}}$ by conjugation. For the definition of a *Cartan involution* see [12], §1.

On the other hand, if G is any adjoint, semisimple algebraic group over $\mathbb{R}$ and $u : \mathbb{U} \to G$ is a homomorphism satisfying the above three properties, then the $G(\mathbb{R})^+$-conjugacy class of u naturally has the structure of a Hermitian symmetric domain D, for which

$$G(\mathbb{R})^+ = \mathrm{Hol}(D)^+$$

and $u(-1)$ is the involution associated to u when regarded as a point of D.

4 The Deligne torus

Let $\mathbb{S}$ denote the linear algebraic group over $\mathbb{R}$ such that $\mathbb{S}(\mathbb{R}) = \mathbb{C}^{\times}$. Similar to the case of $\mathbb{U}$ we may realise the elements of $\mathbb{S}(\mathbb{R})$ as those

$$\begin{pmatrix} a & b \\ c & d \end{pmatrix} \in GL_2(\mathbb{R})$$

such that $a = d$, $b = -c$. This is also a torus, usually referred to as the *Deligne torus*, and we have a short exact sequence

$$1 \to \mathbb{G}_m \xrightarrow{w} \mathbb{S} \to \mathbb{U} \to 1,$$

which on real points corresponds to

$$1 \to \mathbb{R}^{\times} \xrightarrow{r \mapsto r^{-1}} \mathbb{C}^{\times} \xrightarrow{z \mapsto z/\bar{z}} \mathbb{U}(\mathbb{R}) \to 1.$$

Therefore, any homomorphism $u : \mathbb{U} \to G$ yields a homomorphism

$$h : \mathbb{S} \to G,$$

defined by $h(z) = u(z/\bar{z})$. Furthermore, $\mathbb{U}(\mathbb{R})$ will act on $\mathfrak{g}_{\mathbb{C}}$ via the characters $z \mapsto 1$, $z \mapsto z$ and $z \mapsto z^{-1}$ if and only if $\mathbb{S}(\mathbb{R})$ acts on $\mathfrak{g}_{\mathbb{C}}$ via the characters $z \mapsto 1$, $z \mapsto z/\bar{z}$ and $z \mapsto \bar{z}/z$.

Conversely, let $h : \mathbb{S} \to G$ be a homomorphism such that $\mathbb{S}$ acts on $\mathfrak{g}_{\mathbb{C}}$ via the characters $z \mapsto 1$, $z \mapsto z/\bar{z}$ and $z \mapsto \bar{z}/z$. Then $w(\mathbb{G}_m(\mathbb{R}))$ acts trivially on $\mathfrak{g}_{\mathbb{C}}$, which implies that h is trivial on $w(\mathbb{G}_m(\mathbb{R}))$, since the adjoint representation of G on $\mathfrak{g}$ is faithful. Thus, h arises from a homomorphism $u : \mathbb{U} \to G$.

Therefore, to give a $G(\mathbb{R})^+$-conjugacy class D of homomorphisms $u : \mathbb{U} \to G$ satisfying the above three properties is the same as to give a $G(\mathbb{R})^+$-conjugacy class X^+ of homomorphisms $h : \mathbb{S} \to G$ satisfying the following:

- Only the characters $z \mapsto 1$, $z \mapsto z/\bar{z}$ and $z \mapsto \bar{z}/z$ occur in the representation of $\mathbb{S}(\mathbb{R})$ on $\mathfrak{g}_{\mathbb{C}}$.
- Conjugation by $h(i)$ constitutes a Cartan involution of G.
- The element $h(i)$ maps to a non-trivial element in every simple factor of G.

5 Hodge structures

Therefore, the question should be *why are we interested in such conjugacy classes of morphisms $h : \mathbb{S} \to G$?* To understand this, we require the notion of a Hodge structure. Below is a brief summary of the relevant definitions. For a more comprehensive account, we refer the reader to [12], §2.

For a real vector space V, we define complex conjugation on

$$V(\mathbb{C}) := V \otimes_{\mathbb{R}} \mathbb{C}$$

by $\overline{v \otimes z} := v \otimes \bar{z}$. A *Hodge decomposition* of V is a decomposition

$$V(\mathbb{C}) = \bigoplus_{(p,q)\in\mathbb{Z}\times\mathbb{Z}} V^{p,q}$$

such that $\overline{V^{p,q}} = V^{q,p}$. A *Hodge structure* is a real vector space V with a Hodge decomposition. The set of pairs (p,q) such that $V^{p,q} \neq 0$ is called the *type* of the Hodge structure and we refer to a Hodge structure of type $(-1,0), (0,-1)$ as a *complex structure*.

For each $n \in \mathbb{Z}$,

$$\bigoplus_{p+q=n} V^{p,q}$$

is stable under complex conjugation and equal to $V_n(\mathbb{C})$ for some real subspace V_n of V. The decomposition $V = \oplus_n V_n$ is called the *weight decomposition* of V. If $V = V_n$, then V is said to have *weight n*. The *Hodge filtration* associated with a Hodge structure V of weight n is

$$F := \{\cdots \supset F^p \supset F^{p+1} \supset \cdots\}, \quad F^p := \oplus_{r\geq p} V^{r,n-r}.$$

A $\mathbb{Z}$-(respectively $\mathbb{Q}$-)*Hodge structure* is a free $\mathbb{Z}$-module (respectively $\mathbb{Q}$-vector space) V of finite rank (respectively dimension) equipped with a Hodge decomposition of

$$V(\mathbb{R}) := V \otimes \mathbb{R}$$

such that the weight decomposition is defined over $\mathbb{Q}$.

Recall that we can identify $\mathbb{S}$ with a closed subgroup of GL_2 as follows: for any $\mathbb{R}$-algebra A, we realise $\mathbb{S}(A)$ as those matrices of the form

$$\begin{pmatrix} a & b \\ -b & a \end{pmatrix} \in \mathrm{GL}_2(A).$$

Diagonalising, $\mathbb{S}_{\mathbb{C}}$ is isomorphic to $\mathbb{G}_m^2$, with complex conjugation on $\mathbb{S}(\mathbb{C})$ corresponding to $(z_1, z_2) \mapsto (\overline{z_2}, \overline{z_1})$. Therefore, the elements of $\mathbb{S}(\mathbb{R})$ map to the elements $(z, \bar{z})$, stable under conjugation. More generally, the characters of $\mathbb{S}_{\mathbb{C}}$ are the homomorphisms

$$(z_1, z_2) \mapsto z_1^p z_2^q,$$

for any $(p,q) \in \mathbb{Z} \times \mathbb{Z}$, with complex conjugation acting as $(p,q) \mapsto (q,p)$.

Consequently, to give a representation of $\mathbb{S}$ on a real vector space V is the same as to give a $\mathbb{Z} \times \mathbb{Z}$-grading of $V(\mathbb{C})$ such that $\overline{V^{p,q}} = V^{q,p}$ for all p and q, which is precisely the definition of a Hodge structure on V. We thus define morphisms, tensor products and duals of Hodge structures as morphisms, tensor products and duals of representations of $\mathbb{S}$. We normalise the relation so that (z_1, z_2) acts on $V^{p,q}$ as $z_1^{-p} z_2^{-q}$. A complex structure on a real vector space V is then precisely a Hodge structure $\mathbb{S} \to \mathrm{GL}(V)$ coming from a homomorphism $\mathbb{C} \to \mathrm{End}(V)$.

For $n \in \mathbb{Z}$ and $R = \mathbb{Z}$, $\mathbb{Q}$ or $\mathbb{R}$, we let $R(n)$ be the $(R\text{-})$Hodge structure $V = R$, where $\mathbb{S}$ acts on $V(\mathbb{R}) = \mathbb{R}$ by the character $(z\bar{z})^n$ and, hence,

$$V(\mathbb{C}) = V_{-n}(\mathbb{C}).$$

This is referred to as a *Tate twist*. For an $(R\text{-})$Hodge structure V of weight n, a *Hodge tensor* is a multilinear form $t : V^r \to R$ such that the map

$$V \otimes V \otimes \cdots \otimes V \to R(-nr/2)$$

is a morphism of Hodge structures.

If we denote by $C := h(i)$ the *Weil operator*, then a *polarisation* on V is a Hodge tensor

$$\psi : V \times V \to R$$

such that

$$\psi_C : V(\mathbb{R}) \times V(\mathbb{R}) \to \mathbb{R} : (x, y) \mapsto \psi(x, Cy)$$

is symmetric and positive definite. A *polarisation* on an $(R\text{-})$Hodge structure $V = \oplus_n V_n$ is a system $(\psi_n)_n$ of polarisations on the V_n.

6 Abelian varieties

Consider an *Abelian variety A* over $\mathbb{C}$ of dimension g. Then A is isomorphic to a *complex torus* $\mathbb{C}^g / \Lambda$, where Λ is the $\mathbb{Z}$-module generated by an $\mathbb{R}$-basis for $\mathbb{C}^g$. The isomorphism $\Lambda \otimes \mathbb{R} \cong \mathbb{C}^g$ defines a complex structure on $\Lambda \otimes \mathbb{R}$ and there exists an alternating form

$$\psi : \Lambda \times \Lambda \to \mathbb{Z}$$

such that $\psi_{\mathbb{R}}(x, Cy)$ is symmetric and positive definite and

$$\psi_{\mathbb{R}}(Cx, Cy) = \psi_{\mathbb{R}}(x, y),$$

for all $x, y \in \Lambda \otimes \mathbb{R}$. In other words, $\Lambda \cong H_1(A, \mathbb{Z})$ is a $\mathbb{Z}$-Hodge structure of weight -1 equipped with a polarisation. In fact, by [12], Theorem 6.8, the functor $A \mapsto H_1(A, \mathbb{Z})$ is an equivalence from the category of Abelian varieties over $\mathbb{C}$ to the category of polarised $\mathbb{Z}$-Hodge structures of type $(-1, 0), (0, -1)$. Therefore, the answer to the question of the previous section is that *one can study the problem of parameterising Abelian varieties in terms of Hodge structures.*

Consider the case of Abelian varieties of dimension one, otherwise known as *elliptic curves*. An elliptic curve over $\mathbb{C}$ is the quotient of $\mathbb{C}$ by a free $\mathbb{Z}$-module Λ of rank 2. Two elliptic curves $\mathbb{C}/\Lambda$ and $\mathbb{C}/\Lambda'$ are isomorphic if and only if $\Lambda' = \alpha \Lambda$ for some $\alpha \in \mathbb{C}^\times$. We summarise the perspective explained in [9]:

Often, when considering elliptic curves, we fix $\mathbb{C}$ and vary Λ. Instead, however, we may fix $\Lambda := \mathbb{Z}^2$ and vary the complex structure on $\mathbb{Z}^2 \otimes \mathbb{R} = \mathbb{R}^2$ i.e. we vary the morphism

$$h : \mathbb{C}^\times \to \mathrm{GL}_2(\mathbb{R})$$

extending to a homomorphism $\mathbb{C} \to \mathrm{M}_2(\mathbb{R})$ of $\mathbb{R}$-algebras. Given such a morphism, we obtain an isomorphism of complex vector spaces $i_h : \mathbb{R}^2 \to \mathbb{C}$ defined by

$$i_h^{-1}(z) = h(z) \cdot i_h^{-1}(1) := h(z) \cdot e_0,$$

where we choose $e_0 = (1, 0) \in \mathbb{R}^2$. The quotient $\mathbb{C}/i_h(\mathbb{Z}^2)$ is an elliptic curve.

Therefore, let

$$h_0 : \mathbb{C}^\times \to \mathrm{GL}_2(\mathbb{R}) : a + ib \mapsto \begin{pmatrix} a & b \\ -b & a \end{pmatrix}$$

and let $h := \gamma h_0 \gamma^{-1}$, where

$$\gamma = \begin{pmatrix} x & y \\ w & z \end{pmatrix} \in \mathrm{GL}_2(\mathbb{R})^+.$$

Note that, for any such h, the standard symplectic form given by

$$(u, v) \mapsto u^t \begin{pmatrix} 0 & -1 \\ 1 & 0 \end{pmatrix} v$$

is a polarisation for the corresponding $\mathbb{Z}$-Hodge structure.

For $h_0(z)$, the z-eigenspace in $\mathbb{R}^2 \otimes \mathbb{C}$ is the complex subspace generated by $(-i, 1)$. The $\bar{z}$-eigenspace is its complex conjugate, generated by $(i, 1)$. Therefore, for $h(z)$, the z-eigenspace is generated by

$$\begin{pmatrix} x & y \\ w & z \end{pmatrix}\begin{pmatrix} -i \\ 1 \end{pmatrix} = \begin{pmatrix} -xi+y \\ -wi+z \end{pmatrix}$$

or, equivalently, $(\overline{\tau}_h, 1)$, where $\tau_h := xi + y/wi + z$, and the $\overline{z}$-eigenspace is generated by $(\tau_h, 1)$. Note that this latter subspace is precisely the middle term in the filtration associated to the $\mathbb{Z}$-Hodge structure given by h.

Now, i_h extends $\mathbb{C}$-linearly to a map

$$i_{h,\mathbb{C}} : \mathbb{R}^2 \otimes \mathbb{C} = \mathbb{C} \cdot \begin{pmatrix} \overline{\tau}_h \\ 1 \end{pmatrix} \oplus \mathbb{C} \cdot \begin{pmatrix} \tau_h \\ 1 \end{pmatrix} \to \mathbb{C}$$

and, since it commutes with the action of $\mathbb{C}$ on both sides, we deduce that $i_{h,\mathbb{C}}$ is the quotient of $\mathbb{R}^2 \otimes \mathbb{C}$ by the $\overline{z}$-eigenspace. Therefore, since $i_h(e_0) = 1$ and $i_h((0,1)) = i_h(-\tau_h e_0 + (\tau_h, 1)) = -\tau_h$,

$$i_h(\mathbb{Z}^2) = \mathbb{Z} \oplus \mathbb{Z}\tau_h.$$

We conclude that $\mathbb{C}/i_h(\mathbb{Z}^2)$ varies over all isomorphism classes of elliptic curves as h varies over the $\mathrm{GL}_2(\mathbb{R})^+$-conjugacy class of h_0. The map $h \mapsto \tau_h$ is a $\mathrm{GL}_2(\mathbb{R})^+$-equivariant bijection between this conjugacy class and $\mathbb{H}$.

For Abelian varieties of dimension g, the situation is similar. We replace $\mathbb{Z}^2$ by $\mathbb{Z}^{2g}$ and fix the standard symplectic form given by

$$-J := \begin{pmatrix} 0 & -\mathrm{id} \\ \mathrm{id} & 0 \end{pmatrix}.$$

We let

$$h_0 : \mathbb{C}^\times \to \mathrm{GL}_{2g}(\mathbb{R}) : a + bi \mapsto a + bJ,$$

which factors through the group

$$\mathrm{GSp}_{2g}(\mathbb{R}) = \{g \in \mathrm{GL}_{2g}(\mathbb{R}) : g^t J g = \nu(g)J\},$$

where $\nu : \mathrm{GSp}_{2g} \to \mathbb{G}_m$ is a homomorphism of linear algebraic groups. The $\mathrm{GSp}_{2g}(\mathbb{R})^+$-conjugacy class of h_0 corresponds to the set of $\mathbb{Z}$-Hodge structures on $\mathbb{Z}^{2g}$ having type $(-1,0), (0,-1)$ for which J induces a polarisation. Using the description of the Hodge filtration, as in the case of elliptic curves, one can identify this set in a $\mathrm{GSp}_{2g}(\mathbb{R})^+$-equivariant manner with a Hermitian symmetric domain

$$\mathbb{H}_g := \{Z = X + iY \in M_{g\times g}(\mathbb{C}) : Z = Z^t, Y > 0\}$$

called the *Siegel upper half-space of genus g*.

7 The Siegel upper half-space

Let us return then to our account of Hodge structures. Having fixed a $g \in \mathbb{N}$, we denote the Hodge structure corresponding to a point $\tau \in \mathbb{H}_g$ by V_τ and we denote the corresponding Hodge filtration by F_τ. For any given $(p,q) \in \mathbb{Z} \times \mathbb{Z}$, the dimension $d(p,q)$ of $V_\tau^{p,q}$ is constant as τ varies over $\mathbb{H}_g$ and we have a continuous map

$$\tau \mapsto [V_\tau^{p,q}] : \mathbb{H}_g \to G_{d(p,q)}(V(\mathbb{C})),$$

from $\mathbb{H}_g$ to the complex, projective variety of $d(p,q)$-dimensional subspaces of $V(\mathbb{C})$.

The subspace dimensions of F_τ are then also constant as τ varies over $\mathbb{H}_g$ and, if we denote by $F_d(V(\mathbb{C}))$ the complex, projective variety parameterising such filtrations of $V(\mathbb{C})$, then the map

$$f : \tau \mapsto [F_\tau] : \mathbb{H}_g \to F_d(V(\mathbb{C}))$$

is holomorphic. In light of these properties, we refer to the set of Hodge structures corresponding to the points of $\mathbb{H}_g$ as a *holomorphic family of Hodge structures*.

Finally, the differential of f at τ is a $\mathbb{C}$-linear map

$$df_\tau : T_\tau \mathbb{H}_g \to T_{[F_\tau]} F_d(V(\mathbb{C}))$$

from the tangent plane of $\mathbb{H}_g$ at τ to the tangent plane of $F_d(V(\mathbb{C}))$ at $[F_\tau]$. By [12], (17), $T_{[F_\tau]} F_d(V(\mathbb{C}))$ is a subset of

$$\bigoplus_p \mathrm{Hom}(F_\tau^p, V(\mathbb{C})/F_\tau^p)$$

but, in this case, the image of df_τ is actually contained in the space

$$\bigoplus_p \mathrm{Hom}(F_\tau^p, F_\tau^{p-1}/F_\tau^p)$$

and we say that this holomorphic family of Hodge structures is a *variation of Hodge structures*.

8 Families of Hodge structures

The above situation can be abstracted as follows: let V be a finite dimensional $\mathbb{R}$-vector space and let T be a finite set of tensors on V, including a nondegenerate bilinear form t_0. Fix an $n \in \mathbb{N}$ and let

$$d : \mathbb{Z} \times \mathbb{Z} \to \mathbb{N}$$

be a symmetric function such that $d(p, q) = 0$ for almost all (p, q), including every (p, q) such that $p + q \neq n$.

Consider the set $S(d, T)$ of Hodge structures on V such that, for all $(p, q) \in \mathbb{Z} \times \mathbb{Z}$,

$$\dim V^{p,q} = d(p, q),$$

every $t \in T$ is a Hodge tensor and t_0 is a polarisation. This is naturally a subspace of

$$\prod_{(p,q):d(p,q)\neq 0} G_{d(p,q)}(V(\mathbb{C})).$$

Therefore, $S(d, T)$ can be given the subspace topology and, by [12], Theorem 2.14, (assuming it is non-empty) any connected component has a unique complex structure such that the corresponding set of Hodge structures constitute a holomorphic family. Furthermore, if such a family is actually a variation of Hodge structures, then the corresponding connected component S^+ has the structure of a Hermitian symmetric domain. In fact, every Hermitian symmetric domain is of the form S^+ for a suitable V, T and d.

9 The algebraic group

Recall the topological space $S(d, T)$ from the previous section and let S^+ be a connected component. Fix a point $h_0 \in S^+$ and let G be the smallest algebraic subgroup of $\mathrm{GL}(V)$ such that

$$h : \mathbb{S} \to \mathrm{GL}(V)$$

factors through G for every $h \in S^+$ i.e. the intersection of all subgroups having this property. As in the proof of [12], Theorem 2.14 (a), for any $g \in G(\mathbb{R})^+$, $gh_0g^{-1} \in S^+$ and, in fact, the map

$$g \mapsto gh_0g^{-1} : G(\mathbb{R})^+ \to S^+$$

is surjective. In other words, S^+ is the $G(\mathbb{R})^+$-conjugacy class of h_0.

10 Shimura data

Motivated by our example of Abelian varieties, we want to consider $\mathbb{Z}$-(or $\mathbb{Q}$-) Hodge structures. This will be achieved by choosing an algebraic group G defined over $\mathbb{Q}$ and embedding this into $\mathrm{GL}(V)$ for some $\mathbb{Q}$-vector space V. The $\mathbb{Z}$-structure will come from the choice of a lattice in V.

Definition 10.1   A Shimura datum is a pair (G, X), where G is a reductive group over $\mathbb{Q}$ and X is a $G(\mathbb{R})$-conjugacy class of morphisms $h : \mathbb{S} \to G_{\mathbb{R}}$ such that, for one (or, equivalently, all) $h \in X$,

- Only the characters $z \mapsto 1, z \mapsto z/\bar{z}$ and $z \mapsto \bar{z}/z$ occur in the representation of $\mathbb{S}$ on the Lie algebra of $G_{\mathbb{C}}^{\mathrm{ad}}$.

- Conjugation by $h(i)$ is a Cartan involution of G^{ad}.

- For every simple factor H of G^{ad}, the map $\mathbb{S} \to H_{\mathbb{R}}$ is not trivial.

By a *reductive* algebraic group we refer to a connected, linear algebraic group with trivial unipotent radical. The *unipotent radical* of a linear algebraic group is the unipotent part of its radical, where its *radical* is the neutral component of its maximal normal, solvable subgroup. The semisimple groups are those linear algebraic groups with trivial radical. In particular, they are reductive.

Now let (G, X) be a Shimura datum. By the first of the axioms above, $\mathbb{G}_m(\mathbb{R}) = \mathbb{R}^\times$, which is naturally a subgroup of $\mathbb{S}(\mathbb{R}) = \mathbb{C}^\times$, acts trivially on $\mathfrak{g}_{\mathbb{C}}$. As the action of G on $\mathfrak{g}$ factors through G^{ad} and the action of G^{ad} is faithful, the image of $\mathbb{R}^\times$ in $G(\mathbb{R})$ must belong to the centre. In particular, the restriction of any $h \in X$ to $\mathbb{G}_m$ is independent of h and we refer to its reciprocal w as the *weight homomorphism* since, for any representation $\rho : G_{\mathbb{R}} \to \mathrm{GL}(V)$, $\rho \circ w$ defines the weight decomposition of the Hodge structure given by $\rho \circ h$ on V.

Now let $\rho : G_{\mathbb{R}} \to \mathrm{GL}(V)$ be a faithful representation. By [12], Proposition 5.9, X has a unique structure of a complex manifold such that the family of Hodge structures induced on V by $\rho \circ h$ as h varies over X is holomorphic. In fact, the first axiom implies that it is a variation of Hodge structures. Therefore, from our earlier discussion of families of Hodge structures, X is a finite disjoint union of Hermitian symmetric domains.

Alternatively, consider a connected component X^+ of X. By [12], Proposition 5.7 (a), we may consider X^+ as a $G^{\mathrm{ad}}(\mathbb{R})^+$-conjugacy class of morphisms $\mathbb{S} \to G_{\mathbb{R}}^{\mathrm{ad}}$. Let $h \in X^+$ and decompose $G_{\mathbb{R}}^{\mathrm{ad}}$ into a product of simple factors H_i so that $h = (h_i)_i$, where h_i is the projection of h to H_i. By [12], Lemma 4.7, if $H_i(\mathbb{R})$ is compact then h_i is trivial. Otherwise, given the conditions satisfied by h, there exists a Hermitian symmetric domain D_i such that $H_i(\mathbb{R})^+$ coincides with $\mathrm{Hol}(D_i)^+$ and D_i is in natural one-to-one correspondence with the $H_i(\mathbb{R})^+$-conjugacy class X_i^+ of h_i. Therefore, the product D of the D_i is a Hermitian symmetric domain on which $G^{\mathrm{ad}}(\mathbb{R})^+$ acts via a surjective homomorphism $G^{\mathrm{ad}}(\mathbb{R})^+ \to \mathrm{Hol}(D)^+$ with compact kernel and there is a natural identification of D with $X^+ = \prod_i X_i^+$.

Definition 10.2 A morphism of Shimura data

$$(G_1, X_1) \to (G_2, X_2)$$

is a morphism $\phi : G_1 \to G_2$ such that, for every $h \in X_1$, $\phi \circ h \in X_2$. If ϕ is a closed immersion, we refer to (G_1, X_1) as a Shimura subdatum.

Definition 10.3 Let (G, X) be a Shimura datum. Let X^{ad} be the $G^{\mathrm{ad}}(\mathbb{R})$-conjugacy class of morphisms $\mathbb{S} \to G_{\mathbb{R}}^{\mathrm{ad}}$ containing the image of X. Then $(G^{\mathrm{ad}}, X^{\mathrm{ad}})$ is a Shimura datum called the adjoint Shimura datum and

$$(G, X) \to (G^{\mathrm{ad}}, X^{\mathrm{ad}})$$

is a morphism of Shimura data.

11 Congruence subgroups

Let G be a reductive subgroup of GL_n defined over $\mathbb{Q}$. We denote by $G(\mathbb{Z})$ the group $G(\mathbb{Q}) \cap \mathrm{GL}_n(\mathbb{Z})$. Recall the following definition, independent of the embedding of G in GL_n:

Definition 11.1 A subgroup Γ of $G(\mathbb{Q})$ is arithmetic if $\Gamma \cap G(\mathbb{Z})$ has finite index in Γ and $G(\mathbb{Z})$ i.e. if Γ and $G(\mathbb{Z})$ are commensurable.

Now suppose that (G, X) is a Shimura datum. We would like to consider the corresponding Hodge structures up to isomorphism and this is the role of the group Γ. We may also wish to distinguish additional structure to that already encoded in the group G. The most obvious such structure is distinguished by the following class of arithmetic subgroups:

Definition 11.2 The principal congruence subgroup of level N is defined as the group

$$\Gamma(N) := \{g \in G(\mathbb{Z}) : g \equiv \mathrm{id} \bmod N\},$$

where the congruence relation is entry-wise.

In the case of Abelian varieties, where $G = \mathrm{GSp}_{2g}$ and we consider the $\mathbb{Z}$-Hodge structure on $\Lambda = H_1(A, \mathbb{Z})$, the group $\Gamma(N)$ also distinguishes between different bases for the N-torsion subgroup $\frac{1}{N}\Lambda/\Lambda$, rather than simply the isomorphism class of Λ along with its polarisation.

Of course, the definition of the principal congruence subgroup depends on the embedding of G in GL_n. Therefore, we define a *congruence subgroup* of $G(\mathbb{Q})$ to be a subgroup containing some $\Gamma(N)$ as a subgroup of finite index. This notion does not depend on the embedding.

12 Adeles

The ring of *finite (rational) adeles* $\mathbb{A}_f$ comprises the elements

$$\alpha = (\alpha_p) \in \prod_p \mathbb{Q}_p$$

such that, for almost all primes p, $\alpha_p \in \mathbb{Z}_p$. It is endowed with the topology for which a basis of open sets are those of the form $\prod_p U_p$, where U_p is open in $\mathbb{Q}_p$, and $U_p = \mathbb{Z}_p$ for almost all p. Similarly, for an algebraic group G, defined over $\mathbb{Q}$, one can choose an embedding into GL_n and define $G(\mathbb{A}_f)$ as those elements

$$g = (g_p)_p \in \prod_p G(\mathbb{Q}_p)$$

such that $g_p \in \mathrm{GL}_n(\mathbb{Z}_p)$ for almost all p. However, this definition of $G(\mathbb{A}_f)$ is independent of the embedding into GL_n and so is the basis of open sets, defined analogously to the above.

By [12], Proposition 4.1, for any compact open subgroup K of $G(\mathbb{A}_f)$, $K \cap G(\mathbb{Q})$ is a congruence subgroup Γ of $G(\mathbb{Q})$ and every congruence subgroup arises this way. Loosely speaking, considering the congruence relation defining Γ prime-by-prime gives rise to K and vice-versa.

Later, we will also need the more general definition of $\mathbb{A}_{E,f}$, the *finite adeles over a number field E*, which we define as $\mathbb{A}_f \otimes E$ or, equivalently, as the ring of elements

$$\alpha = (\alpha_\upsilon) \in \prod_\upsilon E_\upsilon,$$

over all finite places υ of E such that, for almost all υ, $\alpha_\upsilon \in \mathcal{O}_{E_\upsilon}$. The *adele ring* $\mathbb{A}_E$ arises when we include factors for the infinite places of E. Therefore, any $\alpha \in \mathbb{A}_E$ can be written as a pair $(\alpha_\infty, \alpha_f)$, where $\alpha_f \in \mathbb{A}_{E,f}$.

13 Neatness

Let G be an algebraic subgroup of GL_n defined over $\mathbb{Q}$. The following definition is independent of the embedding into GL_n:

Definition 13.1   An element $g \in G(\mathbb{Q})$ is neat if the subgroup of $\overline{\mathbb{Q}}^\times$ generated by its eigenvalues is torsion free.

One says that a congruence subgroup Γ is *neat* if all of its elements are neat. There is also a notion of neatness for compact open subgroups of $G(\mathbb{A}_f)$, for which we refer the reader to [11], 4.1.4. In particular, if K is neat then so is the congruence subgroup $G(\mathbb{Q}) \cap gKg^{-1}$, for any $g \in G(\mathbb{A}_f)$. Every compact open subgroup K of $G(\mathbb{A}_f)$ contains a neat compact open subgroup K' with finite index.

14 Shimura varieties

Finally, we give the definition of a Shimura variety:

Definition 14.1 Let (G, X) be a Shimura datum and let K be a compact open subgroup of $G(\mathbb{A}_f)$. The Shimura variety attached to (G, X) and K is the double coset space

$$\mathrm{Sh}_K(G, X)(\mathbb{C}) := G(\mathbb{Q}) \backslash X \times (G(\mathbb{A}_f)/K).$$

This definition invariably seems abstruse at first. However, it is a simple calculation to see that

$$\mathrm{Sh}_K(G, X)(\mathbb{C}) = \coprod_{g \in \mathcal{C}} \Gamma'_g \backslash X,$$

where $\mathcal{C}$ is a set of representatives for the double coset space $G(\mathbb{Q}) \backslash G(\mathbb{A}_f)/K$ and $\Gamma'_g := G(\mathbb{Q}) \cap gKg^{-1}$ is a congruence subgroup. Note that, by [21], Theorem 5.1, $\mathcal{C}$ is a finite set. However, since we are interested in connected components, choose a connected component X^+ of X and denote by $G(\mathbb{Q})_+$ its stabiliser in $G(\mathbb{Q})$. Then

$$\mathrm{Sh}_K(G, X)(\mathbb{C}) = \coprod_{g \in \mathcal{C}_+} \Gamma_g \backslash X^+,$$

where $\mathcal{C}_+$ is a set of representatives for the double coset space $G(\mathbb{Q})_+ \backslash G(\mathbb{A}_f)/K$ and $\Gamma_g := G(\mathbb{Q})_+ \cap gKg^{-1}$. By [12], Lemma 5.12, $\mathcal{C}_+$ is also a finite set.

15 Complex structure

Any arithmetic subgroup Γ of $G(\mathbb{Q})$ acts on X through $G^{\mathrm{ad}}(\mathbb{Q})$ and, by [12], Proposition 3.2, its image is also arithmetic. For any arithmetic subgroup Γ of $G(\mathbb{Q})$, the intersection $\Gamma \cap G(\mathbb{Q})_+$ acts on X^+. We say that its image under the map $G^{\mathrm{ad}}(\mathbb{R})^+ \to \mathrm{Hol}(X^+)^+$ is an *arithmetic subgroup* of $\mathrm{Hol}(X^+)^+$.

If Γ is neat then the image of $\Gamma \cap G(\mathbb{Q})_+$ in $\mathrm{Hol}(X^+)^+$ is neat and, in particular, torsion free. By [12], Proposition 3.1, such an arithmetic subgroup of $\mathrm{Hol}(X^+)^+$ acts freely on X^+ and the corresponding quotient has a unique complex structure such that the quotient map is a local isomorphism. In general then, $\Gamma \backslash X^+$ has the structure of a (possibly singular) complex analytic variety.

16 Algebraic structure

The fundamental result of Baily and Borel [1] states that the quotient of X^+ by any torsion free, arithmetic subgroup of $\mathrm{Hol}(X^+)^+$ has a canonical realisation as a complex, quasi-projective, algebraic variety. In particular, if K is neat, $\mathrm{Sh}_K(G,X)(\mathbb{C})$ is the analytification of a quasi-projective variety $\mathrm{Sh}_K(G,X)_{\mathbb{C}}$.

A further theorem of Borel [3] states that, for any smooth, quasi-projective variety V over $\mathbb{C}$, any holomorphic map from $V(\mathbb{C})$ to $\mathrm{Sh}_K(G,X)(\mathbb{C})$ is regular. For example, given any inclusion $K_1 \subset K_2$ of neat compact open subgroups of $G(\mathbb{A}_f)$, we have a natural morphism of algebraic varieties

$$\mathrm{Sh}_{K_1}(G,X)_{\mathbb{C}} \to \mathrm{Sh}_{K_2}(G,X)_{\mathbb{C}}.$$

Therefore, varying K, we get an inverse system of algebraic varieties

$$(\mathrm{Sh}_K(G,X)_{\mathbb{C}})_K$$

and we write the scheme-theoretic limit of this system as $\mathrm{Sh}(G,X)_{\mathbb{C}}$. On the system there is a natural action of $G(\mathbb{A}_f)$ given by

$$\cdot g : \mathrm{Sh}_K(G,X)(\mathbb{C}) \to \mathrm{Sh}_{g^{-1}Kg}(G,X)(\mathbb{C}) : [x,a]_K \mapsto [x,ag]_{g^{-1}Kg},$$

where we use $[\cdot,\cdot]_K$ to denote a double coset belonging to $\mathrm{Sh}_K(G,X)(\mathbb{C})$. By the theorem of Borel, this action is regular on components. Therefore, for any given $g \in G(\mathbb{A}_f)$, we obtain an algebraic correspondence

$$\mathrm{Sh}_K(G,X)_{\mathbb{C}} \leftarrow \mathrm{Sh}_{K \cap gKg^{-1}}(G,X)_{\mathbb{C}} \xrightarrow{\cdot g} \mathrm{Sh}_{g^{-1}Kg \cap K}(G,X)_{\mathbb{C}} \to \mathrm{Sh}_K(G,X)_{\mathbb{C}},$$

where the outer maps are the natural projections. We refer to this correspondence as a *Hecke correspondence*.

Finally, if we have a morphism

$$f : (G_1, X_1) \to (G_2, X_2)$$

of Shimura data and two compact open subgroups $K_1 \subset G_1(\mathbb{A}_f)$ and $K_2 \subset G_2(\mathbb{A}_f)$ such that $f(K_1) \subset K_2$, then we obtain a morphism

$$\mathrm{Sh}_{K_1}(G_1, X_1)(\mathbb{C}) \to \mathrm{Sh}_{K_2}(G_2, X_2)(\mathbb{C}),$$

which, again by the theorem of Borel, is a regular map

$$\mathrm{Sh}_{K_1}(G_1, X_1)_{\mathbb{C}} \to \mathrm{Sh}_{K_2}(G_2, X_2)_{\mathbb{C}}.$$

We refer to the images of such maps as *Shimura subvarieties*. We also have an induced morphism

$$\mathrm{Sh}(G_1, X_1)_{\mathbb{C}} \to \mathrm{Sh}(G_2, X_2)_{\mathbb{C}}$$

of the limits, by which we mean an inverse system of regular maps, compatible with the actions of $G_1(\mathbb{A}_f)$ and $G_2(\mathbb{A}_f)$.

17 Special subvarieties

Special subvarieties constitute the smallest class of irreducible algebraic subvarieties containing the connected components of Shimura subvarieties and closed under taking irreducible components of images under Hecke correspondences. The precise definition is the following:

Definition 17.1   Let $\mathrm{Sh}_K(G,X)_{\mathbb{C}}$ be a Shimura variety. A closed irreducible subvariety Z is called special if there exists a morphism of Shimura data

$$(G',X') \to (G,X)$$

and $g \in G(\mathbb{A}_f)$ such that Z is an irreducible component of the image of

$$\mathrm{Sh}(G',X')_{\mathbb{C}} \to \mathrm{Sh}(G,X)_{\mathbb{C}} \xrightarrow{\cdot g} \mathrm{Sh}(G,X)_{\mathbb{C}} \to \mathrm{Sh}_K(G,X)_{\mathbb{C}}.$$

The situation is analogous to the case of Abelian varieties, where the special subvarieties are the Abelian subvarieties and their translates under torsion points.

By definition, if we let $K' \subset G(\mathbb{A}_f)$ be a compact open subgroup contained in K and consider the natural morphism of Shimura varieties

$$\pi : \mathrm{Sh}_{K'}(G,X)_{\mathbb{C}} \to \mathrm{Sh}_K(G,X)_{\mathbb{C}},$$

- if Z is a special subvariety of $\mathrm{Sh}_{K'}(G,X)_{\mathbb{C}}$, then $\pi(Z)$ is a special subvariety of $\mathrm{Sh}_K(G,X)_{\mathbb{C}}$,
- if Z is a special subvariety of $\mathrm{Sh}_K(G,X)_{\mathbb{C}}$, then any irreducible component of $\pi^{-1}Z$ is a special subvariety of $\mathrm{Sh}_{K'}(G,X)_{\mathbb{C}}$.

18 Special points

The natural definition of a *special point* is then the following:

Definition 18.1   A special point in $\mathrm{Sh}_K(G,X)_{\mathbb{C}}$ is a special subvariety of dimension zero.

However, we can characterise special points in a more concrete manner: consider a special point $[h,g]_K \in \mathrm{Sh}_K(G,X)(\mathbb{C})$. Let $M := \mathrm{MT}(h)$ be the *Mumford-Tate group* of h i.e. the smallest algebraic subgroup H of G (defined over $\mathbb{Q}$) such that $h : \mathbb{S} \to G_{\mathbb{R}}$ factors through $H_{\mathbb{R}}$ and let X_M denote the orbit

$M(\mathbb{R})\cdot h$ inside X. Then (M,X_M) is a Shimura subdatum of (G,X) and, if we let X_M^+ be the connected component $M(\mathbb{R})^+\cdot h$ of X_M, then the image of $X_M^+\times\{g\}$ in $\mathrm{Sh}_K(G,X)(\mathbb{C})$ defines the smallest special subvariety containing $[h,g]_K$. Therefore, X_M must be zero dimensional and so M must be commutative. It is a general fact that any subgroup of G defined over $\mathbb{Q}$ and containing $h(\mathbb{S})$ is reductive. Therefore, M is a torus.

On the other hand if T is a torus in G and $h\in X$ factors through $T_{\mathbb{R}}$ then $[h,g]_K\in\mathrm{Sh}_K(G,X)(\mathbb{C})$ is clearly a special point for any $g\in G(\mathbb{A}_f)$. Therefore, we may define a special point as any point $[h,g]_K\in\mathrm{Sh}_K(G,X)(\mathbb{C})$ such that $\mathrm{MT}(h)$ is a torus. Of course, the choice of h is only well-defined up to conjugation by an element of $G(\mathbb{Q})$, but this doesn't affect the property of $\mathrm{MT}(h)$ being a torus.

19 Canonical model

It is possible to define a model for $\mathrm{Sh}_K(G,X)_{\mathbb{C}}$ that is canonical in a sense one can make precise. As we have seen, $\mathrm{Sh}_K(G,X)(\mathbb{C})$ is often a moduli space for Abelian varieties and the main theorem of complex multiplication gives us a description of how Galois groups act on sets of CM-Abelian varieties. Therefore, we would like the Galois action on $\mathrm{Sh}_K(G,X)(\mathbb{C})$ to agree with this description, whenever it applies. In order to achieve this, the canonical model satisfies a generalised version of this description given in terms of Deligne's group-theoretic (G,X) language. We provide a very brief summary of the theory explained more thoroughly in [12], §12, §13 and §14.

Recall that a *model* over a number field E for a complex algebraic variety V is a variety V_0 defined over E with an isomorphism $\phi:V_{0,\mathbb{C}}\to V$, though we will follow convention and omit any mention of this isomorphism. First we define the field of definition $E:=E(G,X)$ of the canonical model. It is referred to as the *reflex field* and, as we will see, it does not depend on K. This independence is one reason for having several connected components in the definition of a Shimura variety.

For a subfield k of $\mathbb{C}$, we write $\mathcal{C}(k)$ for the set of $G(k)$-conjugacy classes of cocharacters of G_k defined over k i.e.

$$\mathcal{C}(k)=G(k)\backslash\mathrm{Hom}(\mathbb{G}_{m,k},G_k).$$

Any homomorphism $k\to k'$ induces a map $\mathcal{C}(k)\to\mathcal{C}(k')$, so $\mathrm{Aut}(k'/k)$ acts on $\mathcal{C}(k')$.

For $h\in X$, we obtain a cocharacter

$$\mu_h:\mathbb{G}_{m,\mathbb{C}}\xrightarrow{z\mapsto(z,1)}\mathbb{G}_{m,\mathbb{C}}^2\cong\mathbb{S}_{\mathbb{C}}\xrightarrow{h_{\mathbb{C}}}G_{\mathbb{C}}$$

of $G_{\mathbb{C}}$ and so the $G(\mathbb{R})$-conjugacy class X of h maps to an element $c(X) \in \mathcal{C}(\mathbb{C})$. The reflex field E is then the fixed field of the stabiliser of $c(X)$ in $\text{Aut}(\mathbb{C})$. By what follows, we will see that E is a number field.

Suppose that

$$[h, g]_K \in \text{Sh}_K(G, X)(\mathbb{C})$$

is a special point i.e. $M := \text{MT}(h)$ is a torus. Therefore, since all cocharacters of M are defined over $\overline{\mathbb{Q}}$ and μ_h factors through $M_{\mathbb{C}}$, μ_h is defined over a finite extension E_h of $\mathbb{Q}$. Note that E_h does not depend on the choice of h. By [12], Remark 12.3 (b), E is contained in E_h.

For any $t \in M(E_h)$, the element

$$\prod_{\sigma : E_h \to \overline{\mathbb{Q}}} \sigma(t)$$

is stable under $\text{Gal}(\overline{\mathbb{Q}}/\mathbb{Q})$ and so belongs to $M(\mathbb{Q})$. The so-called *reciprocity morphism* is defined by

$$r_h : \mathbb{A}_{E_h,f}^{\times} \to M(\mathbb{A}_f) : a \mapsto \prod_{\sigma : E_h \to \overline{\mathbb{Q}}} \sigma(\mu_h(a)).$$

Finally, recall the (surjective) *Artin map*

$$\text{Art}_{E_h} : \mathbb{A}_{E_h}^{\times} \to \text{Gal}(E_h^{\text{ab}}/E_h)$$

from class field theory and let $\text{Art}_{E_h}^{-1}$ denote its reciprocal.

Definition 19.1 We say that a model of $\text{Sh}_K(G, X)_{\mathbb{C}}$ over E is canonical if every special point $[h, g]_K$ in $\text{Sh}_K(G, X)(\mathbb{C})$ has coordinates in E_h^{ab} and

$$\sigma[h, g]_K = [h, r_h(s_f)a]_K,$$

for any $\sigma \in \text{Gal}(E_h^{\text{ab}}/E_h)$ and $s = (s_\infty, s_f) \in \mathbb{A}_{E_h}^{\times}$ such that $\text{Art}_{E_h}^{-1}(s) = \sigma$.

By [12], Theorem 13.7, if a canonical model exists, it is unique up to unique isomorphism. The difficult theorem is that canonical models actually exist. For a discussion, see [12], §14.

A *model* of $\text{Sh}(G, X)_{\mathbb{C}}$ over E is an inverse system of varieties over E, endowed with a right action of $G(\mathbb{A}_f)$, which over $\mathbb{C}$ is isomorphic to $\text{Sh}(G, X)_{\mathbb{C}}$ with its $G(\mathbb{A}_f)$ action. Such a system is *canonical* if each component is canonical in the above previous sense.

By [12], Theorem 13.7 (b), if for all compact open subgroups K of $G(\mathbb{A}_f)$ $\text{Sh}_K(G, X)_{\mathbb{C}}$ has a canonical model, then so does $\text{Sh}(G, X)_{\mathbb{C}}$ and it is unique up to unique isomorphism. In particular, by [12], Theorem 13.6, the action of $G(\mathbb{A}_f)$ is defined over E. By [12], Remark 13.8, if $(G', X') \to (G, X)$ is a morphism of Shimura data and $\text{Sh}(G',X')_{\mathbb{C}}$ and $\text{Sh}(G,X)_{\mathbb{C}}$ have canonical models, then the induced morphism

$$\text{Sh}(G',X')_{\mathbb{C}} \to \text{Sh}(G,X)_{\mathbb{C}}$$

is defined over $E(G',X') \cdot E(G,X)$.

20 The André-Oort conjecture

The André-Oort conjecture is the following statement regarding the geometry of Shimura varieties:

Conjecture 20.1 Let (G,X) be a Shimura datum, K a compact open subgroup of $G(\mathbb{A}_f)$ and Σ a set of special points in $\text{Sh}_K(G,X)(\mathbb{C})$. Then every irreducible component of the Zariski closure of $\cup_{s\in\Sigma}s$ in $\text{Sh}_K(G,X)_{\mathbb{C}}$ is a special subvariety.

In the remainder of this article, we are going to apply the Pila-Zannier strategy to the André-Oort conjecture. The André-Oort conjecture is analogous to the Manin-Mumford conjecture (first proved by Raynaud [22]), asserting that the irreducible components of the Zariski closure of a set of torsion points in an Abelian variety are the translates of Abelian subvarieties by torsion points. The task at hand is essentially to combine a number of different ingredients. We follow the outline given by Ullmo in [24], §5 for the case of $\mathcal{A}_6^r$.

21 Reductions

Let Y denote an irreducible component of the Zariski closure of $\cup_{s\in\Sigma}s$ in $\text{Sh}_K(G,X)_{\mathbb{C}}$. Let $[h,g]_K \in Y$ denote a point such that $M := \text{MT}(h)$ is maximal among such groups. Note that the maximality is independent of the choice of h. We say that such a point is *Hodge generic* in Y.

Let $X_M := M(\mathbb{R}) \cdot h$. Then, by [8], Proposition 2.1, Y is contained in the image of the morphisms

$$\text{Sh}_{K_M}(M,X_M)_{\mathbb{C}} \to \text{Sh}_{gKg^{-1}}(G,X)_{\mathbb{C}} \xrightarrow{\cdot g} \text{Sh}_K(G,X)_{\mathbb{C}},$$

where $K_M := M(\mathbb{A}_f) \cap gKg^{-1}$. Denote by f their composition and let Y_M be an irreducible component of $f^{-1}Y$. Then Y is a special subvariety of $\text{Sh}_K(G,X)_{\mathbb{C}}$ if and only if Y_M is a special subvariety of $\text{Sh}_{K_M}(M,X_M)_{\mathbb{C}}$. Furthermore, Y_M is Hodge generic in $\text{Sh}_{K_M}(M,X_M)_{\mathbb{C}}$. Therefore, we may assume that Y is Hodge generic in $\text{Sh}_K(G,X)_{\mathbb{C}}$.

Let (G^{ad}, X^{ad}) be the adjoint Shimura datum associated to (G, X) and let K^{ad} be a compact open subgroup of $G^{ad}(\mathbb{A}_f)$ containing the image of K. Then Y is a special subvariety of $\text{Sh}_K(G, X)_{\mathbb{C}}$ if and only if its image Y^{ad} in $\text{Sh}_{K^{ad}}(G^{ad}, X^{ad})_{\mathbb{C}}$ is a special subvariety. Furthermore, if Y is Hodge generic in $\text{Sh}_K(G, X)_{\mathbb{C}}$, then Y^{ad} is Hodge generic in $\text{Sh}_{K^{ad}}(G^{ad}, X^{ad})_{\mathbb{C}}$. Therefore, we may assume that G is semisimple of adjoint type.

Recall that the irreducible components of the image of a special subvariety under a Hecke correspondence are again special subvarieties. Therefore, if we fix a connected component X^+ of X, we may assume that Y is contained in the image $S := \Gamma \backslash X^+$ of $X^+ \times \{1\}$ in $\text{Sh}_K(G, X)(\mathbb{C})$, where $\Gamma := G(\mathbb{Q})_+ \cap K$. We denote a point in S as $[h]$ for some $h \in X^+$.

22 Galois orbits

The first ingredient is a lower bound for the size of the Galois orbit of a special point. By the definition of special subvarieties, the choice of K is irrelevant in the André-Oort conjecture. Thus, we may assume that K is neat and a product of compact open subgroups K_p in $G(\mathbb{Q}_p)$.

Now let $[h] \in S$ be a special point. Recall that $M := \text{MT}(h)$ is a torus and let L denote its *splitting field*, by which we mean the smallest field over which M becomes isomorphic to a product of the multiplicative group. Note that this is a finite, Galois extension of $\mathbb{Q}$ containing E_h and is independent of the choice of h.

Let K_M denote the compact open subgroup $M(\mathbb{A}_f) \cap K$ of $M(\mathbb{A}_f)$, which is equal to the product of the $K_{M,p} := M(\mathbb{Q}_p) \cap K_p$. Let K_M^m be the maximal compact open subgroup of $M(\mathbb{A}_f)$, which is unique since M is a torus and equal to the product of the maximal compact open subgroups $K_{M,p}^m$ of $M(\mathbb{Q}_p)$. Note that $K_{M,p} = K_{M,p}^m$ for almost all primes p. The following conjecture is a natural generalisation of [7], Problem 14, posed by Edixhoven for $\mathcal{A}_g$:

Conjecture 22.1 There exist positive constants c_1, B_1, μ_1 and μ_2 such that, for any special point $[h] \in S$,

$$|\text{Gal}(\overline{\mathbb{Q}}/L) \cdot [h]| > c_1 B_1^{i(M)} [K_M^m : K_M]^{\mu_1} D_L^{\mu_2},$$

where $i(M)$ is the number of places such that $K_{M,p} \neq K_{M,p}^m$ and D_L is the absolute value of the discriminant of L.

Note that, although the groups K_H^m and K_M depend on the choice of h, they are well-defined up to conjugation by an element of Γ and, hence, the index $[K_M^m : K_M]$ is well-defined. By [26], Théorème 6.1, this bound is known to hold under the generalised Riemann hypothesis for CM fields and, by [23], Theorem 1.1, it holds unconditionally in the case of $\mathcal{A}_g$, for g at most 6.

23 Realisations

We refer to a point $h \in X^+$ as a *pre-special point* if $[h] \in S$ is a special point. The second ingredient in the Pila-Zannier strategy is an upper bound for the height of a pre-special point in a fundamental domain $\mathcal{F}$ of X^+ with respect to Γ. As opposed to the case of an Abelian variety, this is a non-trivial issue.

For a sensible notion of height, we must first choose a *realisation* $\mathcal{X}$ of X^+. By this we mean an analytic subset of a complex, quasi-projective variety $\widetilde{\mathcal{X}}$, with a transitive holomorphic action of $G(\mathbb{R})^+$ on $\mathcal{X}$ such that, for any $x_0 \in \mathcal{X}$, the orbit map

$$G(\mathbb{R})^+ \to \mathcal{X} : g \mapsto g \cdot x_0$$

is semi-algebraic and identifies $\mathcal{X}$ with $G(\mathbb{R})^+/K_\infty$, where K_∞ is a maximal compact subgroup of $G(\mathbb{R})^+$ (recall that G is semisimple and adjoint). A morphism of realisations is then a $G(\mathbb{R})^+$-equivariant biholomorphism. By [24], Lemme 2.1, any realisation has a canonical semi-algebraic structure and any morphism of realisations is semi-algebraic. Therefore, X^+ has a canonical semi-algebraic structure.

A subset $Z \subset \mathcal{X}$ is called an *irreducible algebraic subvariety* of $\mathcal{X}$ if Z is an irreducible component of the analytic set $\mathcal{X} \cap \widetilde{Z}$, where $\widetilde{Z}$ is an algebraic subset of $\widetilde{\mathcal{X}}$. By [24], Lemme 2.1, $\mathcal{X} \cap \widetilde{Z}$ has finitely many analytic components and they are semi-algebraic. Also note that, by [10], Corollary B.1, this notion is independent of our choice of $\mathcal{X}$. In particular, we have a well-defined notion of an *irreducible algebraic subvariety* of X^+.

24 Heights

For the remainder of this article, we will fix as our realisation the so-called *Borel embedding* of X^+ into its *compact dual* $X^\vee$. We refer to [27], 3.3 for the following definitions:

As before, for a point $h \in X^+$, let

$$\mu_h : \mathbb{G}_{m,\mathbb{C}} \xrightarrow{z \mapsto (z,1)} \mathbb{G}_{m,\mathbb{C}}^2 \cong \mathbb{S}_{\mathbb{C}} \xrightarrow{h_{\mathbb{C}}} G_{\mathbb{C}}$$

be the corresponding cocharacter and let M_X be the $G(\mathbb{C})$-conjugacy class of μ_h. Let V be a faithful representation of G on a finite dimensional $\mathbb{Q}$-vector space so that, for each point $h \in X^+$, we obtain a Hodge structure V_h and a Hodge filtration

$$F_h := \{ \cdots \supset F_h^p \supset F_h^{p+1} \supset \cdots \}, \; F_h^p := \oplus_{r \geq p} V_h^{r,s}.$$

Fix a point $h_0 \in X^+$ and let P be the *parabolic subgroup* of $G(\mathbb{C})$ stabilising F_{h_0}. We define $X^{\vee}$ to be the complex, projective variety $G(\mathbb{C})/P$, which is naturally a subvariety of the flag variety $\Theta_{\mathbb{C}} := \mathrm{GL}(V_{\mathbb{C}})/Q$, where Q is the parabolic subgroup of $\mathrm{GL}(V_{\mathbb{C}})$ stabilising F_{h_0}. Therefore, we have a surjective map from M_X to $X^{\vee}$ sending μ_h to F_h.

The Borel embedding $X \hookrightarrow X^{\vee}$ is the map $h \mapsto F_h$. It is injective since, by [12], §2, (18), the Hodge filtration determines the Hodge decomposition. In other words, the maximal compact subgroup K_{∞} of $G(\mathbb{R})^+$ constituting the stabiliser of h_0 is equal to $G(\mathbb{R})^+ \cap P$.

However, $\Theta_{\mathbb{C}}$ has a natural model Θ over $\mathbb{Q}$ such that, for any extension L of $\mathbb{Q}$, a point of $\Theta(L)$ corresponds to a filtration defined over L. By definition, $X^{\vee}$ is defined over the reflex field $E := E(G, X)$ and, by the proof of [27], Proposition 3.7, a special point $h \in X^+$ is defined over the splitting field of a maximal torus T of $\mathrm{GL}(V)$ such that $T_{\mathbb{C}}$ contains the Mumford-Tate group of h.

Therefore, since a pre-special point $h \in X^+$ has algebraic coordinates, we are allowed to talk about its *(multiplicative) height* $H(h)$, as defined in [2], Definition 1.5.4. The following theorem due to Orr and the author is a natural generalisation of [17], Theorem 3.1, due to Pila and Tsimerman:

Theorem 24.1 *For any $B_2 > 0$, there exist positive constants c_2, μ_3 and μ_4 such that, for any pre-special point $h \in \mathcal{F}$,*

$$H(h) < c_2 B_2^{i(M)} [K_M^m : K_M]^{\mu_3} D_L^{\mu_4}.$$

Finally, let $h \in X^+$ be a pre-special point and let L be the splitting field of a maximal torus T of $\mathrm{GL}(V)$ such that $T_{\mathbb{C}}$ contains the Mumford-Tate group of h. The dimension d of T is at most the dimension of V and the Galois action on the character group of T is given by a homomorphism

$$\mathrm{Gal}(L/\mathbb{Q}) \hookrightarrow \mathrm{GL}_d(\mathbb{Z}).$$

Since, by a classical result of Minkowski, the number of isomorphism classes of finite groups contained in $\mathrm{GL}_d(\mathbb{Z})$ is finite, the degree of L is bounded by a positive constant depending only on G.

25 Definability

In order to apply the Pila-Wilkie counting theorem, one requires the following theorem:

Theorem 25.1 *The restriction $\pi_{|\mathcal{F}}$ of the uniformisation map*

$$\pi : X^+ \to S$$

is definable in $\mathbb{R}_{\mathrm{an,exp}}$.

This theorem is discussed in several articles. It was first proved for restricted theta functions by Peterzil and Starchenko [14]. In particular, this addressed the case of $\mathcal{A}_g$. It is known for general Shimura varieties due to the work of Klingler, Ullmo and Yafaev [10].

26 Ax-Lindemann-Weierstrass

The final ingredient is the hyperbolic Ax-Lindemann-Weierstrass conjecture. In order to state the conjecture, we require the notion of a *weakly special subvariety*:

Definition 26.1 A variety V in S is weakly special if the (analytic) connected components of $\pi^{-1}V$ are algebraic in X^+.

This definition is actually the characterisation [27], Theorem 1.2 of the original definition [27], Definition 2.1. However, given some familiarity with Shimura varieties, the proof is fairly straightforward and this characterisation is precisely what we need. The term *weakly special* is motivated by the fact that all special subvarieties are weakly special whereas, as explained in [13], weakly special subvarieties are special subvarieties if and only if they contain a special point.

Theorem 26.2 *Let Z be an algebraic subvariety of S. Maximal, irreducible, algebraic subvarieties of $\pi^{-1}Z$ are precisely the irreducible components of the preimages of maximal, weakly special subvarieties contained in Z.*

Again, this problem and its history are discussed at length in several other articles. The theorem above is due to Klingler, Ullmo and Yafaev [10]. It was first proven for compact Shimura varieties by Ullmo and Yafaev [25] and for $\mathcal{A}_g$ by Pila and Tsimerman [18].

27 Pila-Wilkie

Let $A \subset \mathbb{R}^m$ be a definable set in an o-minimal structure and let A^{alg} be the union of all connected, positive dimensional, semi-algebraic subsets contained in A. Recall the Pila-Wilkie counting theorem, first proved for rational points in [19] and later for algebraic points in [15]:

Theorem 27.1 *For every $\epsilon > 0$ and $k \in \mathbb{N}$, there exists a positive constant c, depending only on A, k and ϵ, such that, for any real number $T \geq 1$, the number of points lying on $A \setminus A^{\mathrm{alg}}$, whose coordinates in $\mathbb{R}^m$ are algebraic of degree at most k and of multiplicative height at most T, is at most cT^{ϵ}.*

In this article, the o-minimal structure will be $\mathbb{R}_{\mathrm{an,exp}}$ and *definable* will always mean definable in $\mathbb{R}_{\mathrm{an,exp}}$.

28 Final reduction

The final reduction is the following result due to Ullmo, appearing as Theorem 4.1 in [24]:

Theorem 28.1 *Let Z be a Hodge generic subvariety of $\mathrm{Sh}_K(G,X)_{\mathbb{C}}$, strictly contained in S. Suppose that, if S is a product $S_1 \times S_2$ of connected components of Shimura varieties, then Z is not of the form $S_1 \times Z'$, for a subvariety Z' of S_2. Then the union of all positive dimensional, weakly special subvarieties of $\mathrm{Sh}_K(G,X)_{\mathbb{C}}$ contained in Z is not Zariski dense in Z.*

We apply the theorem to Y noting that the assumption in the theorem is no loss of generality: if necessary, we simply replace S by S_2 and Y by Y'. Thus, we may assume that the union of all positive dimensional special subvarieties of $\mathrm{Sh}_K(G,X)_{\mathbb{C}}$ contained in Y is not Zariski dense in Y.

Therefore, if we are able to show that all but a finite number of special points in Y lie on a positive dimensional special subvariety of $\mathrm{Sh}_K(G,X)_{\mathbb{C}}$ contained in Y, then the theorem implies that $Y = S$.

29 The Pila-Zannier strategy

By Theorem 25.1, $\pi_{|\mathcal{F}}$ is definable and so

$$\widetilde{Y} := \pi^{-1}Y \cap \mathcal{F}$$

is a definable set. By assumption, Y contains a dense set of special points and so is defined over a finite extension F of E.

Consider a pre-special point $h \in \widetilde{Y}$ and let L denote the splitting field of $M := \mathrm{MT}(h)$. The Galois orbit $\mathrm{Gal}(\overline{\mathbb{Q}}/LF) \cdot [h]$ is contained in Y and, if Conjecture 22.1 holds, then

$$|\mathrm{Gal}(\overline{\mathbb{Q}}/LF) \cdot [h]| > c_1' B_1^{i(M)} [K_M^m : K_M]^{\mu_1} D_L^{\mu_2},$$

where $c_1' := c_1/[F : E]$. On the other hand, by [12], Example 12.4 (a), $\mathrm{Gal}(\overline{\mathbb{Q}}/LF) \cdot [h]$ is contained in the image of the morphism

$$\mathrm{Sh}_{K_M}(M, h)(\mathbb{C}) \to \mathrm{Sh}_K(G, X)(\mathbb{C}),$$

induced by the inclusion of Shimura data. Therefore, let

$$[h, m]_K \in \mathrm{Sh}_K(G, X)(\mathbb{C})$$

denote an element of $\mathrm{Gal}(\overline{\mathbb{Q}}/LF) \cdot [h]$, where $m \in M(\mathbb{A}_f)$ is given by the explicit description of the Galois action. Since $[h, m]_K \in S$, m is equal to qk, for some $q \in G(\mathbb{Q})$ and $k \in K$. Denote by h' the point of $\widetilde{Y}$ such that $[h'] = [h, m]_K$. Then, up to conjugation by an element of Γ,

$$M' := \mathrm{MT}(q^{-1} \cdot h) = q^{-1} M q$$

is equal to $\mathrm{MT}(h')$ and

$$K_{M'}^m / K_{M'} = q^{-1} K_M^m q / q^{-1} M(\mathbb{A}_f) q \cap K.$$

Conjugation by q yields a bijection between this quotient and

$$K_M^m / M(\mathbb{A}_f) \cap qKq^{-1},$$

which has cardinality $[K_M^m : K_M]$ since $q = mk^{-1}$.

Consequently, by Theorem 24.1, for $0 < B_2 < \min\{1, B_1\}$ there exist positive constants c_2, μ_3 and μ_4 such that

$$H(h') < c_2 B_2^{i(M)} [K_M^m : K_M]^{\mu_3} D_L^{\mu_4}.$$

Therefore, since all pre-special points in X^+ have algebraic co-ordinates of bounded degree, Theorem 27.1 implies that, for any $\epsilon > 0$, there exists a constant c, depending only on $\widetilde{Y}$ and ϵ, such that there are at most

$$c(c_2 B_2^{i(M)} [K_M^m : K_M]^{\mu_3} D_L^{\mu_4})^{\epsilon}$$

pre-special points on $\widetilde{Y} \setminus \widetilde{Y}^{\mathrm{alg}}$ belonging to $\mathrm{Gal}(\overline{\mathbb{Q}}/LF) \cdot [h]$.

Therefore, we may choose ϵ sufficiently small such that, if either $[K_M^m : K_M]$ or D_L is large enough, then there exists a point in $\mathrm{Gal}(\overline{\mathbb{Q}}/LF) \cdot [h]$ such that the corresponding point $h' \in \tilde{Y}$ belongs to a positive dimensional, semi-algebraic set contained in $\tilde{Y}$. Therefore, by [10], Lemma B.2, h' belongs to an irreducible algebraic subvariety of X^+ contained in $\tilde{Y}$ and so, by Theorem 26.2 (the hyperbolic Ax-Lindemann-Weierstrass theorem), there exists a weakly special subvariety V contained in Y such that $[h'] \in V$. Therefore, V is a special subvariety of positive dimension and $[h]$ belongs to a special subvariety contained in Y.

Therefore, on Y, in the complement of all positive dimensional, special subvarieties contained in Y, the quantities $[K_M^m : K_M]$ and D_L corresponding to special points are bounded. By [28], Proposition 3.21, the set of tori equal to the Mumford-Tate group of a pre-special point such that $[K_M^m : K_M]$ and D_L are bounded lie in only finitely many Γ-conjugacy classes. In particular, such pre-special points lie above only finitely many points in S.

References

[1] *W.L. Baily* and *A. Borel*, Compactification of arithmetic quotients of bounded symmetric domains, Annals of Math., **84** (1966), 442–528

[2] *E. Bombieri* and *W. Gubler*, Heights in Diophantine Geometry, Cambridge University Press (2006)

[3] *A. Borel*, Some metric properties of arithmetic quotients of symmetric spaces and an extension theorem, J. Differential Geometry, **6** (1972), 543–560

[4] *P. Deligne*, Travaux de Shimura, Séminaire Bourbaki, Exposé 389, Fevrier 1971, Lecture Notes in Maths. **244**, Springer-Verlag, Berlin (1971), 123–165

[5] *P. Deligne*, Variétés de Shimura: interprétation modulaire, et techniques de construction de modèles canoniques, Automorphic forms, representations and L-functions (Proc. Sympos. Pure Math., Oregon State Univ., Corvalis, Ore.), Part 2, Proc. Sympos. Pure Math., **XXXIII**, Amer. Math. Soc., Providence, R.I. (1977), 247–289

[6] *M. Demazure* and *A. Grothendieck*, Schémas en groupes, SGA 3, Exp. VIII–XIV, Fasc. 4 (1963–1964), IHES

[7] *B. Edixhoven, B. Moonen* and *F. Oort*, Open problems in algebraic geometry, Bull. Sci. Math., **125** (2001), 1–22

[8] *B. Edixhoven* and *A. Yafaev*, Subvarieties of Shimura varieties, Annals Math., **157** (2003), 621–645

[9] *M. Harris*, Courbes modulaires, available at http://www.math.jussieu.fr/ harris/courses.html

[10] *B. Klingler, E. Ullmo* and *A. Yafaev*, The hyperbolic Ax-Lindemann-Weierstrass conjecture, available at http://www.math.u-psud.fr/ ullmo/ (2013)

[11] *B. Klingler* and *A. Yafaev*, The André-Oort conjecture, Annals Math., to appear

[12] *J. Milne*, Introduction to Shimura varieties, available at www.jmilne.org/math (2004)

[13] *B. Moonen*, Linearity properties of Shimura varieties I, J. Algebraic Geom., **7** (1998), 539–567

[14] *K. Peterzil* and *S. Starchenko*, Definability of restricted theta functions and families of abelian varieties, Duke Math. J., **162** (2013), 627–823

[15] *J. Pila*, On the algebraic points of a definable set, Selecta Math., **15** (2009), 151–170

[16] *J. Pila*, Rational points of definable sets and results of André-Oort-Manin-Mumford type, IMRN, **13** (2009), 2476–2507

[17] *J. Pila* and *J. Tsimerman*, The Andre-Oort conjecture for the moduli space of Abelian surfaces, Compositio Math., **149** (2013), 204–216

[18] *J. Pila* and *J. Tsimerman*, Ax-Lindemann for $\mathcal{A}_g$, Annals Math., to appear

[19] *J. Pila* and *A. Wilkie* The rational points of a definable set, Duke Math. J., **133** (2006), 591–616

[20] *J. Pila* and *U. Zannier*, Rational points in periodic analytic sets and the Manin-Mumford conjecture, Rend. Mat. Acc. Lincei, **19** (2008), 149–162

[21] *V. Platonov* and *A. Rapinchuk*, Algebraic groups and number theory, Pure and Applied Math., **139** (1994), Academic Press, Inc., MA

[22] *M. Raynaud*, Sous-variétés d'une variété abélienne et points de torsion, Arithmetic and Geometry, Vol. I, Birkhauser (1983)

[23] *J. Tsimerman*, Brauer-Siegel for arithmetic tori and lower bounds for Galois orbits of special points, J. Amer. Math. Soc., **25** (2012), 1091–1117

[24] *E. Ullmo*, Quelques applications du théorème d'Ax Lindemann hyperbolique, Compositio Mathematicae, to appear

[25] *E. Ullmo* and *A. Yafaev*, The hyperbolic Ax Lindemann theorem in the compact case, Duke Math Journal, to appear

[26] *E. Ullmo* and *A. Yafaev*, Nombre de classes des tores de multiplication complexe et bornes inérieues pour orbites Galoisiennes de points spéciaux, available at http://www.math.u-psud.fr/ ullmo/ (2013)

[27] *E. Ullmo* and *A. Yafaev*, A characterisation of special subvarieties, Mathematika, **57**, No. 2, (2011), 263–273

[28] *E. Ullmo* and *A. Yafaev*, Galois orbits and equidistribution of special subvarieties: towards the André-Oort conjecture, Annals Math., to appear

6

Lectures on elimination theory for semialgebraic and subanalytic sets

A.J. Wilkie

During the Fall Semester of 2010 I gave a course of lectures at the University of Illinois at Chicago, repeated at the University of Notre Dame, to the graduate students in Logic, and these are the notes of that course. I am extremely grateful to David Marker and Sergei Starchenko for the invitations, and for their kind hospitality during my visit. Many thanks also to the students for typing up the notes, which had remained in scruffy hand written form since I first gave a version of the course to the Logic Advanced Class in Oxford during the Trinity Term of 1994.

My intention in these lecture notes is to present all the mathematical background required for the proof of the quantifier elimination theorem of Denef and van den Dries for the structure $\mathbb{R}_{an}$ in a language with a function symbol for division. Of course, I also give the proof of the theorem itself and here I experimented with using the model theoretic embedding criterion for quantifier elimination rather than following the original paper. However, I now feel that any improvements are minimal and cosmetic.

The prerequisites are, I hope, just a working knowledge of undergraduate algebra and analysis and an introductory graduate course in model theory. So I present the theory of Noetherian rings up to the Artin-Rees Lemma and the Krull Intersection Theorem on the algebraic side, and the basics of convergent power series and analytic functions up to the Weierstrass Preparation Theorem on the analytic side. The two sides come together in the proof of the deepest mathematical result used by Denef and van den Dries, namely the flatness of the ring of convergent power series in the ring of formal power series. (In fact, only the linear closure (and for just one linear equation) of the former ring in the latter is actually needed, so I do not need to mention the general notion of flatness, and thereby avoid a discussion of tensor products.)

O-Minimality and Diophantine Geometry, ed. G. O. Jones and A. J. Wilkie. Published by Cambridge University Press. © Cambridge University Press 2015.

Given the prerequisites, the text is intended to be understood without further references, so I have not included any. So let me mention now that for the algebra I have used

R. Y. Sharp, 'Steps in Commutative Algebra', LMS Student Texts 19, CUP, 1990, and

H. Matsumura, 'Commutative Ring Theory', Cambridge Studies in Advanced Mathematics 8, CUP, 1986.

For the theory of convergent (and formal) power series I have (slavishly) followed

J. M. Ruiz, 'The Basic Theory of Power Series', Advanced Lectures in Mathematics, Vieweg, 1993.

I acquired the material in the first section over many years and through many texts and lectures. It has its origin in the work of Abraham Robinson and I think my very first source was

G. E. Sacks, 'Saturated Model Theory', Mathematics Lecture Notes Series, W.A. Benjamin, 1972.

Finally, the paper itself is

J. Denef and L. van den Dries. p-Adic and Real Subanalytic Sets, Annals of Mathematics 128 (1988), 79–138.

1 Model Theoretic Generalities

Let $\mathcal{L}$ be any first-order language, and

$\exists_0 = \forall_0 :=$ the class of quantifier free $\mathcal{L}$-formulas,

$\exists_{n+1} := \{\varphi \mid \varphi$ is logically equivalent to $\exists \bar{x} \, \psi$ for some $\psi \in \forall_n\}$,

$\forall_{n+1} := \{\varphi \mid \varphi$ is logically equivalent to $\forall \bar{x} \, \psi$ for some $\psi \in \exists_n\}$.

Each class $\exists_n, \forall_n$ is closed under $\wedge$ and $\vee$, and $\varphi \in \exists_n$ (or $\forall_n$) $\Leftrightarrow \neg\varphi \in \forall_n$ (or $\exists_n$, respectively).

For $\mathfrak{A}$ an $\mathcal{L}$-structure, $\mathcal{L}(\mathfrak{A})$ denotes the language obtained by adding a constant symbol c_a for each $a \in \mathfrak{A}$ (i.e. $a \in \mathrm{dom}(\mathfrak{A})$) and $\mathfrak{A}^+$ denotes the natural expansion of $\mathfrak{A}$ to an $\mathcal{L}(\mathfrak{A})$-structure

$$D_n(\mathfrak{A}) := \big\{\varphi \mid \varphi \text{ a } \forall_n\text{-sentence of } \mathcal{L}(\mathfrak{A}) \text{ such that } \mathfrak{A}^+ \models \varphi\big\}.$$

If $\mathfrak{B}$ is another $\mathcal{L}$-structure, $e : \mathfrak{A} \to_n \mathfrak{B}$ means that

(i) e is an embedding from $\mathfrak{A}$ to $\mathfrak{B}$, and

(ii) for all $\forall_n$ (equivalently, $\exists_n$) formulas $\varphi(\bar{x})$ of $\mathcal{L}$, and for all $\bar{a} \in \mathfrak{A}$,

$$\mathfrak{A} \models \varphi(\bar{a}) \leftrightarrow \mathfrak{B} \models \varphi(e(\bar{a})).$$

Further, $e : \mathfrak{A} \to_\infty \mathfrak{B}$ means $e : \mathfrak{A} \to_n \mathfrak{B}$ for all n, i.e. e is *elementary*.

Suppose $\mathfrak{A}$ is an $\mathcal{L}$-structure and $\mathfrak{B}$ an $\mathcal{L}(\mathfrak{A})$-structure such that $\mathfrak{B} \models D_n(\mathfrak{A})$. Then the map $e : \mathrm{dom}(\mathfrak{A}) \to \mathrm{dom}(\mathfrak{B})$ sending $a \mapsto c_a^{\mathfrak{B}}$ satisfies $e : \mathfrak{A} \to_n \mathfrak{B} \restriction \mathcal{L}$. Conversely, if $\mathfrak{C}$ is an $\mathcal{L}$-structure and $e : \mathfrak{A} \to_n \mathfrak{C}$, then $\langle \mathfrak{C}, e(a) \rangle_{a \in \mathfrak{A}} \models D_n(\mathfrak{A})$.

Definition 1.1   An $\mathcal{L}$-theory T is called *model complete* if $D_0(\mathfrak{A}) \cup T$ is a complete $\mathcal{L}(\mathfrak{A})$-theory for any $\mathfrak{A} \models T$.

Theorem 1.2   *Let T be an $\mathcal{L}$-theory. The following are equivalent:*

(i) T is model complete;

(ii) Whenever $\mathfrak{A}, \mathfrak{B} \models T$ and $e : \mathfrak{A} \to_0 \mathfrak{B}$, then $e : \mathfrak{A} \to_1 \mathfrak{B}$;

(iii) Whenever $\mathfrak{A}, \mathfrak{B} \models T$ and $e : \mathfrak{A} \to_0 \mathfrak{B}$ then $e : \mathfrak{A} \to_\infty \mathfrak{B}$;

(iv) For any formula $\varphi(\bar{x})$ there exists an $\exists_1$ formula $\psi(\bar{x})$ such that

$$T \models \forall \bar{x} \, (\varphi(\bar{x}) \leftrightarrow \psi(\bar{x})).$$

Proof   (i)$\Rightarrow$(iii): Let $\varphi(\bar{x})$ be an $\mathcal{L}$-formula, $\bar{a} \subseteq \mathfrak{A}$, and $e : \mathfrak{A} \to_0 \mathfrak{B}$. Suppose $\mathfrak{A} \models \varphi(\bar{a})$. Now $\mathfrak{A}^+, \langle \mathfrak{B}, e(a) \rangle_{a \in \mathfrak{A}} \models D_0(\mathfrak{A}) \cup T$. Hence, by (i), $\mathfrak{A}^+ \equiv \langle \mathfrak{B}, e(a) \rangle_{a \in \mathfrak{A}}$. Therefore, $\mathfrak{B} \models \varphi(e(\bar{a}))$.

(iii)$\Rightarrow$(ii): Obvious.

(ii)$\Rightarrow$(iii): We prove by induction on $n \geq 1$ that for all $\mathfrak{A}, \mathfrak{B} \models T$,

$$e : \mathfrak{A} \to_0 \mathfrak{B} \Rightarrow e : \mathfrak{A} \to_n \mathfrak{B}.$$

The case $n = 1$ is just (ii). Suppose true for some $n \geq 1$. Suppose $\mathfrak{A}, \mathfrak{B} \models T$ and $e : \mathfrak{A} \to_n \mathfrak{B}$. We want to show $e : \mathfrak{A} \to_{n+1} \mathfrak{B}$.

By replacing $\mathfrak{A}$ by its image we may suppose $e = \mathrm{id}_{\mathfrak{A}}$. Hence we may consider the $\mathcal{L}(\mathfrak{B})$-theory $T^* = D_{n+1}(\mathfrak{A}) \cup D_0(\mathfrak{B}) \cup T$. Suppose T^* had no model. Then $D_{n+1}(\mathfrak{A}) \cup T \models \neg \varphi$ for some $\varphi \in D_0(\mathfrak{B})$. Write φ as $\psi(c_{\bar{a}}, c_{\bar{b}})$ where $\bar{a} \subseteq \mathfrak{A}$, $\bar{b} \subseteq \mathfrak{B} \setminus \mathfrak{A}$, and $\psi(\bar{x}, \bar{y})$ is an $\exists_0$-formula of $\mathcal{L}$. Then $D_{n+1}(\mathfrak{A}) \cup T \models \neg \psi(c_{\bar{a}}, c_{\bar{b}})$, so $D_{n+1}(\mathfrak{A}) \cup T \models \forall \bar{y} \, \neg \psi(c_{\bar{a}}, \bar{y})$. Thus $\mathfrak{A} \models \neg \chi(\bar{a})$ where $\chi(\bar{x}) := \exists \bar{y} \, \psi(\bar{x}, \bar{y})$. However, $\mathfrak{B} \models \chi(\bar{a})$ (take $\bar{y} = \bar{b}$), contradicting $\mathrm{id}_{\mathfrak{A}} : \mathfrak{A} \to_n \mathfrak{B}$.

Therefore, T^* has a model, say $\mathfrak{C}$. Then

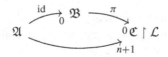

for some π. Hence, by the induction hypothesis,

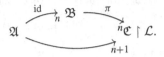

Now suppose $\bar{a} \subseteq \mathfrak{A}$, $\varphi(\bar{x})$ is $\exists_{n+1}$ and $\mathfrak{B} \models \varphi(\bar{a})$. Then $\pi : \mathfrak{B} \to_n \mathfrak{C} \restriction \mathcal{L}$ clearly implies $\mathfrak{C} \restriction \mathcal{L} \models \varphi(\pi(\bar{a}))$. So $\mathfrak{A} \models \varphi(\bar{a})$ since $\pi : \mathfrak{A} \to_{n+1} \mathfrak{C} \restriction \mathcal{L}$, as required.

(iii)$\Rightarrow$(iv): Let $\varphi(\bar{x})$ be any formula of $\mathcal{L}$. Let

$$S = \{\psi(\bar{x}) \mid \psi(\bar{x}) \in \exists_1 \text{ and } T \models \forall \bar{x}\,(\psi(\bar{x}) \to \varphi(\bar{x}))\}.$$

Let $\bar{c}$ be new constant symbols. It suffices to show that

$$T^* = T \cup \{\neg\psi(\bar{c}) \mid \psi(\bar{x}) \in S\} \cup \{\varphi(\bar{c})\}$$

is inconsistent, for then $T \models \forall \bar{x}\left(\bigwedge_{i=1}^{n} \neg\psi_i(\bar{x}) \to \neg\varphi(\bar{x})\right)$ for some $\psi_1, \ldots, \psi_n$ $\in S$ and then $\varphi(\bar{x})$ is equivalent to $\bigvee_{i=1}^{n} \psi_i(\bar{x})$ in T.

Suppose, for a contradiction, that $\mathfrak{A}' \models T^*$. Let $\mathfrak{A} = \mathfrak{A}' \restriction \mathcal{L}$ and $\bar{a} = \bar{c}^{\mathfrak{A}'}$. Then $\mathfrak{A} \models \varphi(\bar{a})$ and $\mathfrak{A} \models \neg\chi(\bar{a})$ for $\chi(\bar{x}) \in S$. Let $T' = T \cup D_0(\mathfrak{A}) \cup \{\neg\varphi(c_{\bar{a}})\}$. Then T' is consistent, for otherwise $T \models \psi(c_{\bar{a}}, c_{\bar{b}}) \to \varphi(c_{\bar{a}})$, for some $\psi(c_{\bar{a}}, c_{\bar{b}}) \in D_0(\mathfrak{A})$ where $\psi(\bar{x}, \bar{y})$ is an $\exists_0$-formula of $\mathcal{L}$, $\bar{b} \subseteq \mathfrak{A} \setminus \{\bar{a}\}$. But then $\exists \bar{y}\,\psi(\bar{x}, \bar{y}) \in S$ and $\mathfrak{A} \models \exists \bar{y}\,\psi(\bar{a}, \bar{y})$, a contradiction. Thus T' has a model, say $\mathfrak{C}$. We have $e : \mathfrak{A} \to_0 \mathfrak{C} \restriction \mathcal{L}$ and $\mathfrak{C} \restriction \mathcal{L} \models \neg\varphi(e(\bar{a}))$, contradicting (iii).

(iv)$\Rightarrow$(i): Suppose $\mathfrak{A} \models T$, $\mathfrak{B} \models T \cup D_0(\mathfrak{A})$. Let $\varphi(c_{\bar{a}})$ be any sentence of $\mathcal{L}(\mathfrak{A})$, where $\bar{a} \subseteq \mathfrak{A}$, $\varphi(\bar{x})$ an $\mathcal{L}$-formula. Let $e : \mathfrak{A} \to \mathfrak{B} \restriction \mathcal{L}$ be the natural embedding. Since $\varphi(\bar{x})$ is T-equivalent to an $\exists_1$-formula by (iv), we have

$$\mathfrak{A}^+ \models \varphi(c_{\bar{a}}) \Rightarrow \mathfrak{B}^+ \models \varphi(c_{\bar{a}}).$$

Hence $\text{Th}(\mathfrak{B}) = \text{Th}(\mathfrak{A}^+)$ for any $\mathfrak{B} \models T \cup D_0(\mathfrak{A})$, and $T \cup D_0(\mathfrak{A})$ is complete. $\square$

Remark   (iv) is the interesting property; it is often proved by showing (ii), which can often be reduced to showing that whenever $\mathfrak{A}, \mathfrak{B} \models T$, $\mathfrak{A} \subseteq \mathfrak{B}$ then $\mathfrak{A}$ is "algebraically closed" in $\mathfrak{B}$ for some natural notion of algebraic closedness.

Definition 1.3   Let T be an $\mathcal{L}$-theory. T is called *substructure complete* if $D_0(\mathfrak{A}) \cup T$ is a complete $\mathcal{L}(\mathfrak{A})$-theory for any $\mathfrak{A}$ which is a substructure of some model of T.

Theorem 1.4 *Let T be an $\mathcal{L}$-theory. The following are equivalent:*

(i) T is substructure complete;

(ii) T eliminates quantifiers, i.e. for any formula $\varphi(\bar{x})$ there exists an $\exists_0$-formula $\psi(\bar{x})$ such that $T \models \forall \bar{x}\, (\varphi(\bar{x}) \leftrightarrow \psi(\bar{x}))$.

Remark If $\bar{x}$ is the empty string here, i.e. if φ is a sentence, then we need to assume that our language contains at least one constant symbol. Where is this needed in the proof below?

Proof (ii)$\Rightarrow$(i): Exercise. Similar to (iv)$\Rightarrow$(i) above.

(i)$\Rightarrow$(ii): Let $\varphi(\bar{x})$ be any $\mathcal{L}$-formula. Let

$$S_0 = \{\psi(\bar{x}) \in \exists_0 \mid T \models \forall \bar{x}\, (\psi(\bar{x}) \to \varphi(\bar{x}))\}.$$

Let $\bar{c}$ be new constant symbols and $T_0 = T \cup \{\varphi(\bar{c})\} \cup \{\neg\psi(c) \mid \psi \in S_0\}$. As in the proof of (iii)$\Rightarrow$(iv) above it is sufficient to show T_0 is inconsistent. Suppose, for contradiction, that $\mathfrak{A} \models T_0$.

Let $S_1 = \{\psi(\bar{x}) \in \exists_0 \mid \mathfrak{A} \models \psi(\bar{c})\}$. Thus $\neg\psi(\bar{x}) \in S_1$ for each $\psi(\bar{x}) \in S_0$. We claim that $T_1 = T \cup \{\neg\varphi(\bar{c})\} \cup \{\psi(\bar{c}) \mid \psi(\bar{x}) \in S_1\}$ has a model. For otherwise, $T \models \forall \bar{x} \left(\bigwedge_{i=1}^{n} \psi_i(\bar{x}) \to \varphi(\bar{x}) \right)$ and hence $\bigwedge_{i=1}^{n} \psi_i(\bar{x}) \in S_0$ for some $\psi_1, \ldots, \psi_n \in S_1$. But then $\bigwedge_{i=1}^{n} \psi_i$ and $\neg \bigwedge_{i=1}^{n} \psi_i$ are in S_1, contradicting the fact that $\mathfrak{A} \models \psi(\bar{c})$ for all $\psi \in S_1$.

Now if $\mathfrak{B} \models T_1$, then $\bar{c}^{\mathfrak{B}}$ and $\bar{c}^{\mathfrak{A}}$ satisfy the same $\exists_0$-formulas in $\mathfrak{B} \upharpoonright \mathcal{L}$ and $\mathfrak{A} \upharpoonright \mathcal{L}$ respectively. Hence they generate canonically isomorphic substructures, and so, by (i), $\mathfrak{A} \equiv \mathfrak{B}$. But $\mathfrak{A} \models \varphi(\bar{c})$ and $\mathfrak{B} \models \neg\varphi(\bar{c})$, a contradiction. $\square$

Theorem 1.5 (Practical test for elimination of quantifiers) *Let T be an $\mathcal{L}$-theory. Suppose that whenever we have $\mathfrak{B}_1, \mathfrak{B}_2 \models T$ and an embedding $e : \mathfrak{A} \to \mathfrak{B}_2$ with $\mathfrak{A} \subseteq \mathfrak{B}_1$, and $\mathfrak{B}_2$ sufficiently saturated (relative to $|\mathfrak{A}|, |\mathcal{L}|$), then for all $a \in \mathfrak{B}_1$, e extends to some $e' : \mathfrak{A}' \to \mathfrak{B}_2$ where $\mathfrak{A} \subseteq \mathfrak{A}' \subseteq \mathfrak{B}$ and $a \in \mathfrak{A}'$. Then T eliminates quantifiers.*

Proof Let $\mathfrak{A} \subseteq \mathfrak{B} \models T$. Let $\mathfrak{B}_1, \mathfrak{B}_2$ be two $\mathcal{L}(\mathfrak{A})$-structures such that $\mathfrak{B}_1, \mathfrak{B}_2 \models T \cup D_0(\mathfrak{A})$. We want to show, by 1.4, that $\mathfrak{B}_1 \equiv \mathfrak{B}_2$. We may suppose $\mathfrak{B}_1, \mathfrak{B}_2$ sufficiently saturated, $\mathfrak{A} \subseteq \mathfrak{B}_1 \upharpoonright \mathcal{L}$. Define $e : \mathfrak{A} \to \mathfrak{B}_2, a \mapsto c_a^{\mathfrak{B}_2}$. One now easily shows by induction on n, using the condition above, that if $e^* : \mathfrak{A}^* \to \mathfrak{B}_2 \upharpoonright \mathcal{L}$ extends e (where $\mathfrak{A} \subseteq \mathfrak{A}^* \subseteq \mathfrak{B}_1 \upharpoonright \mathcal{L}$, $|\mathfrak{A}^*| \leq |\mathfrak{A}| + |\mathcal{L}| + \aleph_0$), then

$$\mathfrak{B}_1 \upharpoonright \mathcal{L} \models \varphi(\bar{a}) \leftrightarrow \mathfrak{B}_2 \upharpoonright \mathcal{L} \models \varphi(e^*(\bar{a}))$$

for all formulas $\varphi(\bar{x})$ of $\mathcal{L}$ containing at most n occurrences of quantifiers and all $\bar{a} \subseteq \mathfrak{A}^*$. Hence $\mathfrak{B}_1 \equiv \mathfrak{B}_2$. $\square$

2 The Real Field

Throughout the rest of this paper we will let $\overline{\mathbb{R}} = \langle \mathbb{R}; +, \cdot, -; 0, 1; < \rangle$ denote the ordered ring of real numbers. We further let $\overline{T} = \mathrm{Th}(\overline{\mathbb{R}})$. We aim to prove that $\overline{T}$ has elimination of quantifiers. (Of course, Tarski's theorem is much stronger: let RCF be the subtheory of $\overline{T}$ axiomatized by the axioms for ordered fields together with $\{\forall y_{n-1}, \ldots, y_0 \exists x \; x^n + y_{n-1}x^{n-1} + \cdots + y_0 = 0 : n = 1, 3, 5, \ldots \}$. Then Tarski showed that RCF has effective elimination of quantifiers and axiomatizes $\overline{T}$.)

Definition 2.1 For k, K, fields with $k \subseteq K$, we say that k is n-closed in K if whenever $p(x) \in k[x]$ is of degree less than or equal to n and $\alpha \in K$ satisfies $p(\alpha) = 0$, then $\alpha \in k$.

Lemma 2.2 *Suppose $K_1, K_2 \models \overline{T}$, $k \subseteq K_1$, k a field, and $e : k \to_0 K_2$. Suppose further that k is n-closed ($n \geq 1$) in K_1 and $e[k]$ is n-closed in K_2. Then if $\alpha \in K_1 \setminus k$ has degree less than or equal to $n + 1$ over k, $\exists e_1 : k[\alpha] \to_0 K_2$ extending e.*

Proof Let $q(x)$ be the minimal monic polynomial of α over k. Then $q(x) = x^{n+1} + a_n x^n + \cdots + a_0$ ($a_0, \ldots, a_n \in k$) is irreducible over k since $\alpha \notin k$ and k is n-closed in K_1. The image, q^e say, of q under e is therefore irreducible over $e[k]$. Choose $a, b \in k$ such that $a < \alpha < b$ (e.g. $a = -\sum_{i=1}^{n} |a_i|$, $b = \sum_{i=1}^{n} |a_i|$). Now, all the roots in K_1 of $q'(x) = 0$ lie in k ($q'(x)$ is the formal derivative of q), so we may suppose that a and b are chosen so that $q'(x)$ has no roots in (a, b). By using sentences in $\overline{T}$ (quantify over the coefficients) it follows that $q(a)$ and $q(b)$ have opposite signs. Therefore, so do $q^e(e(a))$ and $q^e(e(b))$ (since $e : k \to K_2$) and hence (again by using sentences in $\overline{T}$) $q^e(x)$ has a root, say β in K_2 such that $e(a) < \beta < e(b)$. Since $q(x)$ is irreducible, $k(\alpha) \cong e[k](\beta)$ as fields (via $\alpha \mapsto \beta$, $s \mapsto e(s)$ ($s \in k$)).

We must show that this map preserves order. So, suppose that $p(x) \in k[x]$ is any polynomial. Choose $s(x), r(x) \in k[x]$ such that $p(x) \equiv s(x)q(x) + r(x)$ and $\deg r \leq n$ (recall that q is monic of degree n). Then $p^e(x) \equiv s^e(x)q^e(x) + r^e(x)$, $p(\alpha) = r(\alpha)$, $p^e(\beta) = r^e(\beta)$.

Suppose that $p(\alpha) > 0$. Then we want to show $p^e(\beta) > 0$. We have that $r^e(\alpha) > 0$, so it is sufficient to show that $r^e(\beta) > 0$.

Now all roots in K_1 of $r(x)$ lie in k, Similarly for K_2, $r^e(x)$, and $e[k]$. Now choose some $a' \geq a$ and $b' \leq b$ (in k) such that $\alpha \in (a', b')$ and (a', b') contains no roots of $r(x)$. Then $(e(a'), e(b'))$ contains no roots of $r^e(x)$ (consider e^{-1} to see that this is true). Now, $r(\alpha) > 0$, therefore $r(\frac{a'+b'}{2}) > 0$. Thus, $r^e(\frac{e(a')+e(b')}{2}) > 0$, and hence, $r^e(x) > 0$ for all $x \in (e(a'), e(b'))$. Now, with this information, we see that it is sufficient to show that $\beta \in (e(a'), e(b'))$.

Now, q is monotonic on (a', b'), and has a root therein. Hence, $q(a')$ and $q(b')$ have opposite signs. Thus, $q^e(e(a'))$ and $q^e(e(b'))$ have different signs, which implies that q^e has *some* root in $(e(a'), e(b'))$. Suppose it is not β. Then q^e has at least two distinct roots in $(e(a), e(b))$ and so q'^e has a root in $(e(a), e(b))$. Since $\deg q^e \leq n$, this root lies in $e[k]$, and hence (via e^{-1}) q' has a root in (a, b)–contradiction. $\qquad\square$

Lemma 2.3 *Suppose $K_1, K_2 \models \overline{T}$, k_i is a subfield of K_i ($i = 1, 2$), and $e : k_1 \to k_2$ is an isomorphism. Then there exist k_i^* (subfield of K_i) such that $k_i \subseteq k_i^* \subseteq K_i$ ($i = 1, 2$) and an isomorphism $e^* : k_1^* \to k_2^*$ extending e such that k_i^* is n-closed in K_i for all n.*

Proof Just let

$$S = \{< k_1', k_2', e' > : k_i \subseteq k_i' \subseteq K_i, k_i' \text{ a subfield of } K_i \text{ and } e' \text{extends } e\}.$$

Now, order S by extension. Then S satisfies the hypotheses of Zorn's lemma. Hence, S has a maximal element, $\langle k_1^*, k_2^*, e^* \rangle$, say. Then k_i^* is certainly 1-closed in K_i (since k_i^* is a field). Then, 2.2 tells us that if k_i^* is n-closed in K_i, then it is $n + 1$-closed. $\qquad\square$

Lemma 2.4 *Suppose that $K_1, K_2 \models \overline{T}$, k a subring of K_1 and $e : k \to_0 K_2$. Suppose further that K_2 is sufficiently saturated (with respect to $|k|$). Then $\forall \alpha \in K_1$, we can extend e to $e' : k[\alpha] \to_0 K_2$.*

Proof We may clearly extend e to the subfield of K_1 generated by k, so we may as well suppose that k is a subfield of K_1. Now, apply 2.3 with $k_1 = k$ and $k_2 = e[k]$ to get $e^* : k_1^* \to k_2^*$, extending e, an isomorphism with k_i^* n-closed in K_i for all n and $i = 1, 2$. If $\alpha \in k_1^*$, we're already done (let $e' = e^* \upharpoonright k[\alpha]$). Otherwise, α is transcendental over k_1^*. Without loss of generality, we may assume that $\alpha > 0$. Now suppose $p_1(x), \ldots, p_n(x) \in k[x]$. Let $S = \cup_{i=1}^n \{\text{roots of } p_i \text{ in } K_1\}$.

Then $S \subseteq k_1^*$. Let $a = \sup(\{0\} \cup \{\beta \in S : \beta < \alpha\})$. If $\alpha > \beta$ for all $\beta \in S$, then $p_i(a + 1)$ and $p_i(\alpha)$ have the same sign for $i = 1, \ldots, n$. Otherwise, let $b = \inf(\beta \in S : \alpha < \beta)$. Then $p_i(\frac{a+b}{2})$ and $p_i(\alpha)$ have the same sign for

$i = 1, \ldots, n$. In either case, there is some $c \in k_1^*$ such that sign$p_i(c)$ =sign$p_i(\alpha)$ for $i = 1, \ldots, n$. Therefore, sign($p_i^e(e^*(c))$) =sign$p_i(\alpha)$ for $i = 1, \ldots, n$.

Hence,

$$\{p^e(x) > 0 \wedge q^e(x) < 0 : p(x), q(x) \in k[x], K_1 \models p(\alpha) > 0 \wedge q(\alpha) < 0\}$$

is finitely satisfiable in K_2 and therefore satisfiable, by saturation, by $\gamma \in K_2$, say. Clearly, extending e by $\alpha \mapsto \gamma$ gives the required embedding $e' : k[\alpha] \to_0 K_2$. □

Theorem 2.5 $\overline{T}$ *eliminates quantifiers.*

Proof Apply 2.4 and 1.5. □

3 Preliminary Remarks on Rings and Modules

We will take as convention, 'ring' to always mean a commutative ring with identity. Ring homomorphisms preserve 1, and scalar multiplication by 1 is the identity map on any module. Let R be a ring. An R-module is *finite* if it is finitely generated (as an R-module). Recall that R is *Noetherian* if every ideal of R is finitely generated, i.e. if any submodule of the R-module R is finite.

An R-module M is called *Noetherian* if every submodule of M is finite. If R is Noetherian then any finite R-module is as well. For if M is a finite R-module containing a non-finite submodule then by Zorn's lemma, there exists a maximal such, N say. Since M is finite, pick $m \in M \backslash N$ and let N' be the submodule of M generated by N and m. Then N' is finite (by maximality of N), so it is generated by m and $m_1, \ldots, m_n \in N$. (Remark: if something is finitely generated then any generating set contains a finite subset which is generating.) Let $J = \{b \in R : bm \in N\}$. Then J is an ideal of R, so it is finitely generated, say by $b_1, \ldots, b_s$. I claim that the finite subset $\{m_1, \ldots, m_n, b_1 m, \ldots, b_s m\}$ of N generates all of N – contradiction. For suppose that $m' \in N$. Then certainly $m' \in N'$, so $a_1 m_1 + \cdots + a_n m_n + bm = m'$ for some $a_1, \ldots, a_n, b \in R$. It follows that $b \in J$, so $b = r_1 b_1 + \cdots + r_s b_s$ for some $r_1, \ldots, r_s \in R$. We now have

$$a_1 m_1 + \cdots + a_n m_n + r_1(b_1 m) + \cdots + r_s(b_s m) = m',$$

as required. The same proof seems to prove a totally different theorem, I.S. Cohen's lemma:

Lemma 3.1 *A ring R is Noetherian if and only if every prime ideal of R is finitely generated.*

Proof As above, if R contains a non-finitely generated ideal it contains a maximal such, call it I. We show I is prime. Suppose not, say $\alpha\beta \in I$, for $\alpha, \beta \in R \setminus I$. Then the ideal generated by β and I is finitely generated and also $\{b \in R : b\beta \in I\}$ is an ideal of R containing I and α. This ideal is finitely generated. Now proceed as above.   □

We use this to deduce the famous Hilbert Basis Theorem.

Theorem 3.2 *Let R be a Noetherian ring. Then the ring $R[x]$ of polynomials in the variable x over R is also Noetherian.*

Proof Every $f(x) \in R[x]$ may be written in the form $f(0) + xg(x)$ for some $g(x) \in R[x]$ (with $\deg g < \deg f$).

Let P be a prime ideal of $R[x]$. If $x \in P$ then it follows that whenever $f(x) \in P$ then $f(0) \in P$, so P is generated by $\{f(0) : f(x) \in P\} \cup \{x\}$. But $\{f(0) : f(x) \in P\}$ is an ideal of R, so is finitely generated. Thus P is finitely generated as an ideal of $R[x]$.

So, suppose that $x \notin P$. If P is not finitely generated, the following process may be continued indefinitely:

$f_1(x) =$ a polynomial of minimal degree in P,

$f_{n+1}(x) =$ a polynomial of minimal degree in P not in the ideal generated by $\{f_1(x), \ldots, f_n(x)\}$.

Clearly, $f_n(x) \in P$ for all n and $i \leq j$ implies that $\deg f_i \leq \deg f_j$. Now the ideal of R generated by $\{f_n(0) : n = 1, 2, \ldots\}$ is finitely generated by $f_1(0), \ldots, f_N(0)$, say. So there exists $a_1, \ldots, a_N \in R$ such that

$$a_1 f_1(0) + \cdots + a_N f_N(0) - f_{N+1}(0) = 0.$$

Therefore,

$$a_1 f_1(x) + \cdots + a_N f_N(x) - f_{N+1}(x) = xg(x)$$

for some $g(x) \in R[x]$. Now, $\deg g < \deg f_{N+1}$, and $xg(x) \in P$. Hence $g(x) \in P$ since $x \notin P$ and P is prime. It follows that $g(x)$ is in the ideal generated by $f_1(x), \ldots, f_N(x)$ (otherwise we could have chosen $f_{N+1}(x)$ of smaller degree) and hence $f_{N+1}(x)$ is also in this ideal. This is a contradiction.   □

Corollary 3.3 *Let R, S be rings with $R \subseteq S$ and let S be finitely generated (as a ring) over R. Suppose further that R is Noetherian. Then S is also Noetherian.*

Proof Suppose that $s_1, \ldots, s_n$ generates S over R. By 3.2, and induction, the polynomial ring $R[x_1, \ldots, x_n]$ in the variables $x_1, \ldots, x_n$ is Noetherian. There exists an surjective homomorphism $R[x_1, \ldots, x_n] \rightarrow S$ sending $r \mapsto r$ for $r \in R$ and $x_i \mapsto s_i$. But clearly, a homomorphic image of a Noetherian ring is also Noetherian. $\square$

4 Formal Power Series Rings

Definition 4.1 Let R be a ring. The ring $R[[x]]$ of *formal power series* in the variable x consists of, by definition, all series of the form

$$f(x) = \sum_{i=0}^{\infty} a_i x^i \qquad (a_i \in R, i = 0, 1 \ldots)$$

with addition defined by

$$\sum_{i=0}^{\infty} a_i x^i + \sum_{i=0}^{\infty} b_i x^i = \sum_{i=0}^{\infty} (a_i + b_i) x^i$$

and multiplication by

$$\left(\sum_{n=0}^{\infty} a_n x^n \right) \cdot \left(\sum_{n=0}^{\infty} b_n x^n \right) = \sum_{n=0}^{\infty} \left(\sum_{i+j=n} a_i b_j \right) x^n.$$

One readily checks that $R[[x]]$ is a ring and we identify R with the subring of $R[[x]]$ consisting of the f's having $0 = a_1 = a_2 = \ldots$. Write $f(0)$ for a_0 ($f(r)$ has no meaning for any other $r \in R$).

Theorem 4.2 *Suppose that R is Noetherian. Then so is $R[[x]]$.*

Proof Every $f(x) \in R[[x]]$ may be written as $f(0) + xg(x)$ for some $g(x) \in R[[x]]$, so if we let P be a prime ideal of $R[[x]]$, then as in 3.2, if $x \in P$, then P is finitely generated. Suppose $x \notin P$. Then the ideal $\{f(0) : f(x) \in P\}$ of R is finitely generated by $f_1(0), \ldots, f_n(0)$, say. I claim that $f_1(x), \ldots, f_n(x)$ generates P. Let $g(x) \in P$, and suppose that for some $l \geq -1$, we have polynomials $p_1^{(l)}(x), p_2^{(l)}(x), \ldots, p_n^{(l)}(x) \in R[[x]]$ of degree $\leq l$ such that

$$(*)_l : \qquad \sum_{i=1}^{n} p_i^{(l)}(x) f_i(x) - g(x) = x^{l+1} h_l(x)$$

for some $h_l(x) \in R[[x]]$. We set $p_i^{(-1)}(x) \equiv 0$ for $i = 1, \ldots, n$. Then certainly $x^{l+1} h_l(x) \in P$ so $h_l(x) \in P$ since $x \notin P$ and P is prime. Hence there exist

$r_1, \ldots, r_n \in R$ such that $\sum_{i=0}^{n} r_i f_i(0) = h_l(0)$. Then

$$\sum_{i=1}^{n} (p_i^{(l)}(x) - r_i x^{l+1}) f_i(x) - g(x) = x^{l+1} \underbrace{\left(h_l(x) - \sum_{i=1}^{n} r_i f_i(x) \right)}_{s(x)}.$$

But $s(0) = 0$ so $s(x) = x h_{l+1}(x)$ for some $h_{l+1}(x) \in R[[x]]$ and we've 'extended' the $p_i^{(l)}$'s to obtain $(*)_{l+1}$.

Let $\phi_i(x)$ be the unique element of $R[[x]]$ extending all of the $p_i^{(l)}$'s for $i = 1, \ldots, n$. Then clearly the coefficient of x^l in $\sum_{i=1}^{n} \phi_i(x) f_i(x) - g(x)$ is the same as the coefficient of x^l in $\sum_{i=1}^{n} p_i^{(l)}(x) f_i(x) - g(x)$, i.e. 0, for all l. Hence $g(x) = \sum_{i=1}^{n} \phi_i(x) f_i(x)$, and we are done. $\qquad \square$

Theorem 4.3 $f(x)$ *is a unit in* $R[[x]]$ *if and only if* $f(0)$ *is a unit in* R.

Proof   First, suppose that $f(x)$ is a unit in $R[[x]]$. Then, $f(x)g(x) = 1$ implies that $f(0)g(0) = 1$, so $f(0)$ is a unit in R.

For the converse, let

$$f(x) = \sum_{n=0}^{\infty} a_n x^n$$

with a_0 a unit in R. Define $b_0 = a_0^{-1}$ and

$$b_{n+1} = -a_0^{-1}(a_1 b_n + a_2 b_{n-1} + \cdots + a_{n+1} b_0).$$

Then we have

$$\sum_{i+j=0} a_i b_j = 1, \qquad \sum_{i+j=n} a_i b_j = 0 \text{ for } n \geq 1.$$

Hence $\sum_{n=0}^{\infty} a_n x^n \cdot \sum_{n=0}^{\infty} b_n x^n = 1$. $\qquad \square$

Theorem 4.4   *For an ideal I of R, let $\hat{I}$ be the ideal of $R[[x]]$ generated by I and x. Then $I \mapsto \hat{I}$ is a bijection from the maximal ideals of R to those of $R[[x]]$.*

Proof   Proof left as exercise for the reader. $\qquad \square$

5 Adically Normed Rings

Begin by fixing a ring R.

Definition 5.1 A function $|| \cdot || : R \to \mathbb{R}$ is called a *norm* (or rather, an *adic norm*) on R if for all a and b in R, we have:

1. $0 \leq ||a|| \leq 1$ with $||a|| = 0 \leftrightarrow a = 0$;
2. $||ab|| \leq ||a|| \, ||b||$;
3. $||a + b|| \leq \max\{||a||, ||b||\}$.

Now, fix an adic norm $|| \cdot ||$ on R.

Lemma 5.2 *Let $u, a \in R$, with u a unit. Then $||u|| = 1$ and $||ua|| = ||a||$.*

Proof $0 < ||1|| = ||1 \cdot 1|| \leq ||1|| \cdot ||1|| \leq 1$. Hence, $||1|| = 1$. Then, $1 = ||1|| = ||u \cdot u^{-1}|| \leq ||u|| \cdot ||u^{-1}|| \leq 1$. Thus, $||u|| = ||u^{-1}|| = 1$.

Finally, $||ua|| \leq ||u|| \, ||a|| = ||a|| = ||u^{-1}ua|| \leq ||u^{-1}|| \, ||ua|| = ||ua||$. Thus, $||ua|| = ||a||$. $\qquad\square$

Definition 5.3 R is *complete* with respect to $|| \cdot ||$ if $\langle R, d \rangle$ is a complete metric space, where $d(a, b) = ||a - b||$ for $a, b \in R$. Note that $d(a, b) = d(b, a)$ by 5.2.

Remark Note the following facts.

• Suppose that $a \in R$, $||a|| < 1$ and that R is complete with respect to $|| \cdot ||$. Then $1 + a$ is a unit in R.
• Define $|| \cdot ||_0 : R[[x]] \to \mathbb{R}$ by

$$||\sum_{i=0}^{\infty} a_i x^i|| = \sup\{||a_i|| : i = 0, 1, \dots\}.$$

Then, $|| \cdot ||_0$ is an adic norm on $R[[x]]$ and $R[[x]]$ is complete with respect to $|| \cdot ||_0$ if R is complete with respect to $|| \cdot ||$.

Proof Proof left as exercise for the reader. $\qquad\square$

Definition 5.4 Let $p \in \mathbb{N}$, and $\Phi(x) \in R[[x]]$. Then $\Phi(x)$ is called *regular of order p* (with respect to $|| \cdot ||$) if $\Phi(x) = \sum_{i=0}^{p-1} a_i x^i + u(x)x^p$ where $a_0, \dots, a_{p-1} \in R$ satisfy $||a_i|| < 1$ for $i = 0, \dots, p - 1$ and $u(x)$ is a unit of $R[[x]]$.

Theorem 5.5 (Contraction Mapping Theorem) *Let $\langle Y, d \rangle$ be a complete metric space, $c \in \mathbb{R}$, $0 \leq c < 1$, and $T : Y \to Y$ a function such that $d(T(a), T(b)) \leq c \cdot d(a, b) \; \forall a, b \in Y$. Then $\exists! \alpha \in Y$ such that $T(\alpha) = \alpha$.*

Proof Proof left as exercise for the reader. $\qquad\square$

Theorem 5.6 (Division Theorem for $R[[x]]$)   *Assume $\langle R, || \cdot || \rangle$ is a complete normed ring, $p \in \mathbb{N}$, and $\Phi(x) \in R[[x]]$ is regular of order p. Then for any $f(x) \in R[[x]]$, there exist unique $Q(x) \in R[[x]]$ and $S(x) \in R[x]$ with deg $S(x) < p$ such that $f(x) = Q(x) \cdot \Phi(x) + S(x)$.*

Proof   Define $|| \cdot ||_0$ on $R[[x]]$ as in the remark above. Say $\Phi(x) = \sum_{i=0}^{p-1} a_i x^i + u(x)x^p$ as in 5.4. Let $\phi(x) = \sum_{i=0}^{p-1} a_i x^i$ so that $||\phi(x)||_0 = max\{||a_i|| : i = 0, \ldots, p-1\} = c$, say where $0 \le c < 1$.

For any $Q(x) \in R[[x]]$ consider $f(x) - \phi(x)Q(x) \in R[[x]]$. Then $\exists T_Q^*(x) \in R[[x]]$ such that $f(x) - \phi(x)Q(x) = S_Q(x) + x^p T_Q^*(x)$, uniquely where $S_Q(x) \in R[x]$ and has degree $< p$. We want a $Q(x)$ such that $T_Q^*(x) = Q(x)u(x)$.

Define $T : R[[x]] \to R[[x]]$ by $T(Q(x)) = u(x)^{-1} T_Q^*(x)$. We need a $Q(x) \in R[[x]]$ such that $T_Q(x) = Q(x)$.

Now
$$||T(Q_1(x)) - T(Q_2(x))||_0 = ||T_{Q_1}^*(x) - T_{Q_2}^*(x)||_0 \quad \text{(by lemma 5.2)}$$
$$= ||x^p(T_{Q_1}^*(x) - T_{Q_2}^*(x))||_0 \quad \text{(by the definition of } || \cdot ||_0)$$
$$\le ||S_{Q_1}(x) - S_{Q_2}(x) + x^p(T_{Q_1}^*(x) - T_{Q_2}^*(x))||_0$$
$$= ||\phi(x)(Q_2(x) - Q_1(x))||_0 \quad \text{(from the definitions of } S_{Q_i}(x), T_{Q_i}^*(x))$$
$$\le ||\phi(x)||_0 ||Q_1(x) - Q_2(x)||_0$$
$$= c||Q_1(x) - Q_2(x)||_0.$$

Hence, by the Contraction Mapping Theorem, $\exists! Q(x) \in R[[x]]$ such that $T(Q(x)) = Q(x)$. Then $T_Q^*(x) = u(x)q(x)$, so $f(x) - \phi(x)Q(x) = S_Q(x) + x^p u(x)Q(x)$.

Therefore, $f(x) = (\phi(x) + u(x)x^p)Q(x) + S_Q(x) = \Phi(x)Q(x) + S_Q(x)$, as required. $\qquad\square$

Corollary 5.7   *If $\Phi(x) \in R[[x]]$, with $\langle R, || \cdot || \rangle$ complete, is regular of order p, then $\exists b_0, \ldots, b_{p-1} \in R$ with $||b_i|| < 1$ such that $\Phi(x) = v(x)(x^p + \sum_{i=0}^{p-1} b_i x^i)$ where $v(x)$ is a unit of $R[[x]]$.*

Proof   Take $f(x) = x^p$ in theorem 5.6. Then $x^p - S(x) = Q(x)\Phi(x)$ (as in the conclusion of 5.6).

That is, for suitable $b_0, \ldots, b_{p-1} \in R$,

$$x^p + \sum_{i=0}^{p-1} b_i x^i = Q(x)\Phi(x)$$

$$= Q(x)(\sum_{i=0}^{p-1} a_i x^i + u(x)x^p) = (\sum_{i=0}^{\infty} q_i x^i)(\sum_{i=0}^{p-1} a_i x^i + u(x)x^p) \quad \text{(with the } q_i \in R).$$

Equating the coefficients of $1, x, \ldots, x^{p-1}$ shows that $||b_i|| < 1$ for $i = 0, \ldots, p-1$. Equating the coefficients of x^p shows that $q_0 u(0) = 1 + a$ for some $a \in R$ with $||a|| < 1$.

By theorem 4.3, $u(0)$ is a unit of R and hence $Q(x)$ is a unit of $R[[x]]$, as required. $\qquad\qquad\qquad\qquad\qquad\qquad\qquad\qquad\qquad\qquad\qquad\qquad\qquad\qquad\qquad \square$

6 Formal Power Series in Many Variables

Let R be a ring. Define $\mathscr{F}_0(R) = R$ and $\mathscr{F}_{n+1}(R) = \mathscr{F}_n(R)[[x_{n+1}]]$. Here, $x_1, x_2, \ldots, x_{n+1}, \ldots$ are independent variables. $\mathscr{F}_n(R)$ is also written as $R[[x_1, \ldots, x_n]]$. The multi-index notation will be useful for us. For $v \in \mathbb{N}^n$, say $v = \langle v_1, \ldots, v_n \rangle$, write X for $x_1, \ldots, x_n$ and X^v for $x_1^{v_1} \cdots x_n^{v_n}$ (similarly for an n-tuple of elements of R). Also, $|v| := v_1 + \cdots + v_n$.

Exercise 6.1 *We may write the elements of $\mathscr{F}_n(R)$ uniquely in the form* $\sum_{v \in \mathbb{N}^n} a_v X^v$ *($a_v \in R$) so that* $\sum_v a_n X^v + \sum_v b_n X^n = \sum_v (a_n + b_n) X^n$ *and* $\sum_v a_n X^v \cdot \sum_v b_n X^n = \sum_v (\sum_{\lambda + \mu = v} a_\lambda \cdot b_\mu) X^v$ *where addition of multi-indices is co-ordinatewise.*

For $f \in \mathscr{F}_n(R), f(0, \ldots, 0) := a_{0, \ldots, 0}$. Let J be the ideal of $\mathscr{F}_n(R)$ generated by $x_1, \ldots, x_n$. Define a function $ord : \mathscr{F}_n(R) \to \mathbb{N} \cup \{\infty\}$ by

$$ord(f) = \begin{cases} \text{the largest } m \text{ such that } f \in J^m \text{ if such exists,} \\ \infty \text{ otherwise.} \end{cases}$$

(So if $f \notin J$, $ord(f) = 0$ ($J^0 = \mathscr{F}_n(R)$).)

Exercise 6.2 *$ord_n(f) = \infty$ iff $f = 0$ and $ord_n(f) = $ the smallest m such that, writing $f(x) = \sum_v a_v X^v$, we have $a_v \neq 0$ for some v with $|v| = m$. Further, $ord_n(f + g) \geq min(ord_n(f), ord_n(g))$ and $ord_n(f \cdot g) \geq ord_n(f) + ord_n(g)$. It follows that $|| \cdot ||^n$ is a norm on $\mathscr{F}_n(R)$ where $||f||^n = 2^{-ord_n(f)}$ (we set $2^{-\infty} := 0$). Further $|| \cdot ||^n$ is discrete on R (i.e. $||a||^n = 1$ for $a \in R \backslash \{0\}$), and $||f||^n < 1 \Leftrightarrow f \in J \Leftrightarrow f(0, \ldots, 0) = 0$.*

Exercise 6.3 $\langle \mathscr{F}_n(R), || \cdot ||^n \rangle$ *is complete. ($\{f_j\}$ being a Cauchy sequence just means that the f_j become more in agreement as $j \to \infty$.)*

Now consider $f \in \mathscr{F}_n(R) = \mathscr{F}_{n-1}(R)[[x_n]]$.

Then f being regular of order p with respect to $|| \cdot ||_{n-1}$ according to 5.4 means

$$f(x_1, \ldots, x_n) = \sum_{i=0}^{p-1} a_i(x_1, \ldots, x_{n-1})x_n^i + u(x_i, \ldots, x_n)x_n^p$$

where $a_1, \ldots, a_{p-1} \in \mathscr{F}_{n-1}(R)$ and $a_i(0, \ldots, 0) = 0$ (for $i = 0, \ldots, p-1$) and where $u(x_1, \ldots, x_n)$ is a unit of $\mathscr{F}_n(R)$. By 4.3 and induction, the latter is equivalent to $u(0, \ldots, 0)$ being a unit in R which, if R is a field, just means that $u(0, \ldots, 0) \neq 0$. This in turn is equivalent to $a_p(0, \ldots, 0) \neq 0$.

Hence, we obtain from 5.7:

Theorem 6.4 (Weierstrass Preparation Theorem for $\mathscr{F}_n(R)$) *Suppose R is a field and $n \geq 1$. Let $\Phi(x_1, \ldots, x_n) \in \mathscr{F}_n(R)$ with*

$$\Phi(x_1, \ldots, x_n) = \sum_{i=0}^{\infty} a_i(x_1, \ldots, x_{n-1})x_n^i.$$

Suppose Φ is regular in x_n of order p. Then there exists a unique unit $v(x_1, \ldots, x_n) \in \mathscr{F}_n(R)$ and unique $b_0(x_1, \ldots, x_{n-1}), \ldots, b_{p-1}(x_1, \ldots, x_{n-1}) \in \mathscr{F}_{n-1}(R)$, all 0 at $(0, \ldots, 0)$, such that $\Phi(x_1, \ldots, x_n) = v(x_1, \ldots, x_n)(x_n^p + b_{p-1}(x_1, \ldots, x_{n-1})x_n^{p-1} + \cdots + b_0(x_1, \ldots, x_{n-1}))$.

Remark 6.5 Obviously there is a corresponding version of the division theorem for $\mathscr{F}_n(R)$, R a field.

Theorem 6.6 (The Formal Denef-van den Dries Preparation Theorem) *Let R be any Noetherian ring and let $\Phi(X)$ be any element of $\mathscr{F}_n(R)$, $n \geq 1$. Then there are a positive integer d, elements a_v of R, and units $u_v(X)$ of $\mathscr{F}_n(R)$ for $v \in \mathbb{N}^n$ with $|v| < d$ such that $\Phi(X) = \sum_{|v|<d} a_v X^n \cdot u_v(X)$.*

Proof Case $n = 1$.

So $X = x_1$ and $\mathscr{F}_1(R) = R[[X]]$. Say $\Phi(X) = \sum_{i=0}^{\infty} a_i X^i$ (the a_i being in R).

Consider the ideal of R generated by $\{a_0, a_1, \ldots\}$. Since R is Noetherian, it is generated by a_i for $i < d$, for some positive integer d.

Therefore, for each $j \in \mathbb{N}$, we may choose $b_{i,j} \in R$ for $i < d$ such that

$$a_{d+j} = \sum_{i<d} b_{i,j} a_i.$$

Hence, we have

$$\Phi(X) = (\sum_{i<d} a_i X^i) + \sum_{j=0}^{\infty} (\sum_{i<d} b_{i,j} a_i) X^{d+j}$$

$$= (\sum_{i<d} a_i X^i) + X^d \sum_{i<d} a_i (\sum_{j=0}^{\infty} b_{i,j} X^j)$$

$$= \sum_{i<d} a_i X^i (1 + X^{d-i} \sum_{j=0}^{\infty} b_{i,j} X^j).$$

Hence, we are done with $u_i(x) = 1 + X^{d-i}(\sum_{j=0}^{\infty} b_{i,j} X^j)$ ($= 1$ at the origin, so this is a unit).

Now if $\Phi(X) = \Phi(x_1, \ldots, x_{n+1}) \in \mathscr{F}_{n+1}(R) = \mathscr{F}_n(R)[[x_{n+1}]]$, we apply the case $n = 1$ to the variable x_{n+1} and the Noetherian (by 4.2 and induction) ring $\mathscr{F}_n(R)$. So $\Phi(X) = \sum_{i<d} a_i X^i_{n+1} u_i(X)$ with all $a'_i s$ in $\mathscr{F}_n(R)$. Now apply the induction hypothesis to the a_i.   □

6.7 Substitution *Let R be an arbitrary ring. $J_n :=$ ideal of $\mathscr{F}_n(R)$ generated by $x_1, \ldots, x_n$.*

Note that for any $m \geq 0$, J_n^m consists (by definition) of all elements of the form $h(g_1, \ldots, g_r)$ for $r \in \mathbb{N}$, $h(\zeta_1, \ldots, \zeta_r) \in R[\zeta_1, \ldots, \zeta_r]$ homogeneous of degree m, and $g_1, \ldots, g_r \in J_n$.

Now fix r and consider some $f \in \mathscr{F}_r(R)$ and $g_1, \ldots, g_r \in J_n$. Write

$$f(x_1, \ldots, x_r) = \sum_{i=0}^{\infty} h_i(x_1, \ldots, x_r),$$

where h_i is a homogeneous polynomial in $R[x_1, \ldots, x_n]$ of degree i. Then $\forall N, M$

$$(\sum_{i=0}^{N+M} h_i(g_1, \ldots, g_r) - \sum_{i=0}^{N} h_i(g_1, \ldots, g_r)) \in J_n^{N+1},$$

and hence this element of $\mathscr{F}_n(R)$ has $|| \cdot ||^n$−norm $\leq 2^{-(N+1)}$.

Thus, $\{\sum_{i=0}^{N} h_i(g_1, \ldots, g_r)\}_{N \in \mathbb{N}}$ is a Cauchy sequence in $\mathscr{F}_n(R)$ and hence converges by 6.3. We denote its limit by $f(g_1, \ldots, g_r)$.

*Thus, for fixed $g_1, \ldots, g_r \in J_n$, we have a map $\mathscr{F}_r(R) \to \mathscr{F}_n(R)$ given by :
$f \mapsto f(g_1, \ldots, g_r)$.*

Exercise 6.8 *Check that this map is continuous (for $\| \cdot \|^r$, $\| \cdot \|^n$) and a homomorphism. Show further that if $p_1, \ldots, p_n \in J_1$ ($\trianglelefteq \mathscr{F}_1(R)$), then*

$$f(g_1, \ldots, g_r)(p_1, \ldots, p_n) = f(g_1(p_1, \ldots, p_n), \ldots, g_r(p_1, \ldots, p_n)).$$

Now suppose $f \in \mathscr{F}_n(R), f \neq 0$, and $\text{ord}(f) = p$. Then by 6.2, we can write

$$f(x_1, \ldots, x_n) = \sum_{|v| \geq p} a_v X^v \quad (a_v \in R)$$

$$= \sum_{i=p}^{\infty} h_i(x_1, \ldots, x_n),$$

where the h_i are homogeneous polynomials of degree i and $h_p \neq 0$.

Consider $x_1 + c_1 x_n, \ldots, x_{n-1} + c_{n-1} x_n, x_n$, where $c_1, \ldots, c_{n-1} \in R$. These are in J_n.

Substitute :

$$f(x_1 + c_1 x_n, \ldots, x_{n-1} + c_{n-1} x_n, x_n) = \sum_{i=p}^{\infty} h_i(x_1 + c_1 x_n, \ldots, x_{n-1} + c_{n-1} x_n, x_n)$$

$$= \sum_{i=0}^{\infty} b_i(x_1, \ldots, x_{n-1}) x_n^i, \text{ say.}$$

Using 6.8, set $x_1, \ldots, x_{n-1}$ to zero. We have

$$\sum_{i=0}^{\infty} b_i(0, \ldots, 0) x_n^i = \sum_{i=p}^{\infty} h_i(c_1, \ldots, c_{n-1}, 1) x_n^i.$$

So $b_i(0, \ldots, 0) = \begin{cases} 0 & \text{for } i < p \\ h_i(c_1, \ldots, c_{n-1}) & \text{for } i \geq p. \end{cases}$

If R is an infinite field, we can choose $c_1, \ldots, c_{n-1}$ so that $h_p(c_1, \ldots, c_{n-1}, 1) \neq 0$. Hence we have proved the following :

Lemma 6.9 *Suppose R is an infinite field, $f \in \mathscr{F}_n(R)$, $f \neq 0$, $\text{ord}(f) = p$ ($< \infty$). Then after an invertible linear change of co-ordinates, f becomes regular in x_n of order p.*

7 Convergent Power Series

In this section $R = \mathbb{R}$ or $\mathbb{C}$. Write K for R and $\mathcal{F}_n$ for $\mathcal{F}_n(K)$. Recall that for $\alpha = \langle \alpha_1, \cdots, \alpha_n \rangle \in K^n$, and $v = \langle v_1, \ldots, v_n \rangle \in \mathbb{N}^n$, α^v denotes $\alpha_1^{v_1} \cdots \alpha_n^{v_n}$. Also $|\cdot|$ denotes the usual modulus on K.

Definition   For $f = f(X) \in \mathcal{F}_n$, with $n \geq 1$, $X = (X_1, \ldots, X_n)$, say $f(X) = \sum_v a_v X^v$ with $a_v \in K$ for all v, let $\mathrm{dom}(f)$ be the interior of the set of all points $\alpha \in K^n$ such that the set $\{|a_v \alpha^v| : v \in \mathbb{N}^n\}$ is bounded. We say that f is convergent iff $\mathrm{dom}(f) \neq \emptyset$. The set $\{f \in \mathcal{F}_n : f \text{ is convergent}\}$ is denoted O_n.

Example   Let $f(X_1, X_2) = \sum_{i=0}^{\infty} X_1^i X_2^i$. We have $|1 \cdot \alpha_1^i \alpha_2^i|$ is bounded iff $|\alpha_1 \alpha_2| \leq 1$. Thus $\mathrm{dom}(f) = \{\langle \alpha_1, \alpha_2 \rangle \in \mathbb{R}^2 : |\alpha_1 \alpha_2| < 1\}$.

Lemma 7.1   Suppose $f \in \mathcal{F}_n$ is convergent. Then (with $f(X) = \sum_v a_v X^v$)

1. $\mathrm{dom}(f)$ is a non-empty, open subset of K^n;
2. whenever $\alpha \in \mathrm{dom}(f)$, there is a $c \in R^n$ with $0 < c_i < 1$ $(c = \langle c_1, \ldots, c_n \rangle)$ and $M \in \mathbb{R}$ such that $|a_v \cdot \alpha^v| \leq Mc^v$ for all $v \in \mathbb{N}^n$;
3. whenever $\alpha \in \mathrm{dom}(f)$, $\beta \in K^n$ and $|\beta_i| \leq |\alpha_i|$ for $i = 1, \ldots, n$ then $\beta \in \mathrm{dom}(f)$;
4. $\mathrm{dom}(f)$ is a connected subset of K^n;
5. for $\alpha \in \mathrm{dom}(f)$, $\sum_v a_v \alpha^v$ is absolutely convergent.

Proof   (1) Immediate.

(2) Suppose $\alpha \in \mathrm{dom}(f)$. Then $t \cdot \alpha = \langle t_1 \alpha_1, \ldots, t_n \alpha_n \rangle \in \mathrm{dom}(f)$ for some $t = \langle t_1, \ldots, t_n \rangle \in R^n$ with $t_1, \ldots, t_n > 1$ by part (1). Then there is some M such that $|a_v (t \cdot \alpha)^v| \leq M$ for all $v \in \mathbb{N}^n$. . Hence the result with $c_i = t_i^{-1}$.

(3) By (1) we may suppose $|\beta_i| < |\alpha_i|$ for $i = 1, \ldots, n$. Then $|\gamma_i| \leq |\alpha_i|$ for all $\gamma = \langle \gamma_1, \ldots, \gamma_n \rangle$ close to β. Then if $|a_v \alpha^v| \leq M$ we have $|a_v \gamma^v| \leq M$ for all $v \in \mathbb{N}^n$ and γ close to β.

(4) By (3) and the fact that $\mathrm{dom}(f) \neq \emptyset$ we have that $0 \in \mathrm{dom}(f)$, and also that if $\alpha \in \mathrm{dom}(f)$ then the ray from 0 to α is included in $\mathrm{dom}(f)$.

(5) This follows from the following calculation:

$$\sum_{|v| \leq N} |a_v \alpha^v| \leq M \sum_{|v| \leq N} c^v \qquad (\text{from (2)})$$

$$\leq M \prod_{i=1}^{n} \left(\sum_{j=0}^{N} c_i^j \right)$$

$$\leq \frac{M}{(1 - c_1) \cdots (1 - c_n)}. \qquad \square$$

For $f \in \mathcal{F}_n$ convergent and $\alpha \in \mathrm{dom}(f)$ let $\tilde{f}(\alpha)$ be the sum of the series $\sum_\nu a_\nu \alpha^\nu$, where $f = \sum a_\nu X^\nu$. So $\tilde{f} : \mathrm{dom}(f) \to K$, $\mathrm{dom}(f)$ being a connected, open subset of K^n containing 0.

Lemma 7.2 *Suppose $f, g \in \mathcal{F}_n$ are convergent. Then so are $f \pm g, f \cdot g$. Indeed we have $\mathrm{dom}(f * g) \supseteq \mathrm{dom}(f) \cap \mathrm{dom}(g)$ and $\widetilde{f * g} = \tilde{f} * \tilde{g}$, where $* \in \{\pm, \cdot\}$. Thus O_n is a subring of $\mathcal{F}_n$.*

Proof We just do the multiplication case. So suppose that $f(X) = \sum_\nu a_\nu X^\nu$, and $g(X) = \sum_\nu b_\nu X^\nu$. Then

$$f \cdot g = \sum_\nu \left(\sum_{\lambda + \mu = \nu} a_\lambda b_\mu \right) X^\nu.$$

Let $\alpha \in \mathrm{dom}(f) \cap \mathrm{dom}(g)$. By 7.1(2) we may choose $c \in \mathbb{R}^n$ where $c = \langle c_1, \ldots, c_n \rangle$ with $0 < c_i < 1$ for $i = 1, \ldots, n$, and an $M \in \mathbb{R}$ such that $|a_\nu \alpha^\nu|, |b_\nu \alpha^\nu| \leq M c^\nu$ for all ν. Then for $\nu \in \mathbb{N}^n$

$$\left| \left(\sum_{\lambda + \mu = \nu} a_\lambda b_\mu \right) \alpha^\nu \right| = \left| \sum_{\lambda + \mu = \nu} a_\lambda \alpha^\lambda b_\mu \alpha^\mu \right| \leq M^2 c^\nu (\nu_1 + 1) \cdots (\nu_n + 1)$$

which approaches 0 as $|\nu| \to \infty$, and hence is bounded. $\qquad \square$

Lemma 7.3 *Suppose $f \in O_r$ and $g_1, \ldots, g_r \in O_n \cap J_n$. Then $f(g_1, \ldots, g_r) \in O_n$ and*

$$\widetilde{f(g_1, \ldots, g_r)} = \tilde{f}(\tilde{g}_1, \ldots, \tilde{g}_r)$$

sufficiently close to zero.

Proof For a formal power series h let $|h|$ be the formal series obtained by 'modding' the coefficients. Clearly $\mathrm{dom}(|h|) = \mathrm{dom}(h)$. Also, the coefficients of $f(g_1, \ldots, g_r)$ are bounded in absolute value by those of $|f|(|g_1|, \ldots, |g_r|)$. Now $\widetilde{g_i}(0) = 0$ for $i = 1, \ldots, r$ and hence, by continuity, there is an open (box) neighborhood U of zero in K^n such that $U \subseteq \cap_{i=1}^r \mathrm{dom}(g_i)$, and such that $\langle \widetilde{|g_1|}(\alpha), \ldots, \widetilde{|g_r|}(\alpha) \rangle \in \mathrm{dom}(f)$ for all $\alpha \in U$.

Now fix $\alpha \in U$. We want to show that $\alpha \in \mathrm{dom}(f(g_1, \ldots, g_r))$. It is clearly sufficient to show $\alpha \in \mathrm{dom}(|f|(|g_1|, \ldots, |g_r|))$. Fix some $\nu \in \mathbb{N}^n$ and choose a polynomial truncation $|f|_\nu$, say, of $|f|$ such that $|f|(|g_1|, \ldots, |g_r|)$ and $|f|_\nu(|g_1|, \ldots, |g_r|)$ have the same coefficient of X^ν. Now, by 7.2

$$\widetilde{|f|_\nu(|g_1|, \ldots, |g_r|)}(|\alpha|) = \widetilde{|f|_\nu}(\beta_1, \ldots, \beta_r)$$
$$\leq \widetilde{|f|}(\beta_1, \ldots, \beta_r)$$

where $|\alpha| = \langle|\alpha_1|, \ldots, |\alpha_n|\rangle$ and $\beta_i = |g_i|(|\alpha|)$ for $i = 1, \ldots, r$. Note that everything in sight is non-negative, and $\alpha \in U$ implies $|\alpha| \in U$. Let $|f|(|g_1|, \ldots, |g_r|) = \sum_\nu a_\nu X^\nu$. Hence, from the equation above, $|a_\nu \alpha^\nu| = |a_\nu| \cdot |\alpha|^\nu \leq \widetilde{|f|}(\beta_1, \ldots, \beta_r)$ since the middle term is one of the terms in $\widetilde{|f|}_\nu(\beta_1, \ldots, \beta_r)$.

Thus $|a_\nu \alpha^\nu|$ is bounded independently of ν, and so, since α is an arbitrary member of the open set U, $\alpha \in \mathrm{dom}(|f|(|g_1|, \ldots, |g_r|))$ as required.

The second part follows easily, again using approximation to f (and 7.2).

$\qquad\qquad\qquad\qquad\qquad\qquad\qquad\qquad\qquad\qquad\qquad\qquad\qquad\qquad\square$

Corollary 7.4 *Suppose $f \in O_n$ is a unit in $\mathcal{F}_n$ (i.e. $f(0) = \tilde{f}(0) \neq 0$). Then f is a unit in O_n. Hence the ideal generated by $X_1, \ldots, X_n$ is the unique maximal ideal of O_n.*

Proof Suppose $f(0) \neq 0$. Then $f = a + g$ where $g \in O_n$, $g(0) = 0$ and $a \in K^*$. We may suppose $a = 1$ (consider $a^{-1}f$). Let $h(X_1) = \sum_{i=0}^\infty X_1^i$. Then $h \in O_1$ and $(1 - X_1)h(X_1) = 1$. Apply the map $\mathcal{F}_1 \to \mathcal{F}_n$ given by $j \mapsto j(-g)$. We get $(1 + g)h(-g) = 1$ (i.e. $f \cdot h(-g) = 1$).

By 7.3 $h(-g) \in O_n$ since $h \in O_1$ and $-g \in O_n \cap J_n$. $\qquad\qquad\qquad\square$

Exercise 7.5 *Suppose $f \in O_n$, then $f \in \mathcal{F}_n = \mathcal{F}_{n-1}[[X_n]]$. Hence*

$$f(X_1, \ldots, X_n) = \sum_{i=0}^\infty f_i(X_1, \ldots, X_{n-1})X_n^i$$

uniquely with the f_i's in $\mathcal{F}_{n-1}$.

Then each f_i is convergent and $\pi(\mathrm{dom}(f)) \subset \mathrm{dom}(f_i)$ for each i, where π : $(t_1, \ldots, t_n) \mapsto (t_1, \ldots, t_{n-1})$ is the projection onto the first $n-1$ coordinates.

Theorem 7.6 (Division Theorem for O_n) *Suppose $\Phi \in O_n$ is regular of order p with respect to X_n. Then for every $f \in O_n$ there exists a unique $Q \in O_n$ and $S \in O_{n-1}[X_n]$ of degree $< p$ such that $f = Q\Phi + S$.*

Proof For $\alpha = \langle\alpha_1, \ldots, \alpha_n\rangle \in \mathbb{R}^n$ with $\alpha_i > 0$ for $i = 1, \ldots, n$, let

$$O_n^\alpha = \{f \in O_n : |f|(\bar{x}) \text{ converges at } \alpha\}.$$

Then (exercise) $\langle O_n^\alpha, \|\cdot\|\rangle$ is a complete normed space with $\|\sum_\nu a_\nu X^\nu\| = \sum_\nu |a_\nu|\alpha^\nu$. Now we have

$$\Phi(X_1, \ldots, X_n) = \sum_{i=0}^{p-1} f_i(X_1, \ldots, X_{n-1})X_n^i + u(X_1, \ldots, X_n)X_n^p \qquad (6.1)$$

where $f_i \in \mathcal{F}_{n-1}$ for $i = 0, \ldots, p-1$ and u is a unit of $\mathcal{F}_n$. But $\Phi \in O_n$ so $f_i \in O_{n-1}$ by 7.5 and then, clearly, $u \in O_n$ and u is a unit of O_n by 7.4. Further, by hypothesis,

$$f_i(0) = \tilde{f}_i(0) = 0$$

for $i = 0, \ldots, p-1$. Let U be an open box in K^n centered at 0 and contained in the domain of all the series appearing in 6.1 and such that $U \subseteq \operatorname{dom}(f)$ and such that for some $M \geq 1$

$$|\widetilde{||u^{-1}||}(\alpha)| < M$$

for all $\alpha \in U$. Say

$$U = \left\{ \langle \beta_1, \ldots, \beta_n \rangle \in K^n \mid |\beta_i| < \epsilon, i = 1, \ldots, n \right\}.$$

Now set $\alpha_n = \epsilon/2$ and choose $0 < \alpha_1, \ldots, \alpha_{n-1} < \epsilon$ such that

$$\sum_{i=0}^{p-1} |\widetilde{f}_i|(\alpha_1, \ldots, \alpha_{n-1}) \cdot \alpha_n^i < \frac{\alpha_n^p}{2M}.$$

This is possible since $|\widetilde{f}_i|$ is continuous on U and $|\widetilde{f}_i|(0) = 0$. Let $|| \cdot ||$ be the norm described above on O_n^α. Let $\phi(X_1, \ldots, X_n) = \sum_{i=0}^{p-1} f_i(X_1, \ldots, X_{n-1})X_n^i$. Then $f, \Phi, f_1, \ldots, f_{p-1}, \phi, u, u^{-1} \in O_n^\alpha$ and $||\phi|| < \alpha_n^p/2M$, $||u^{-1}|| < M$.

Now for $Q \in O_n^\alpha$ define $T_Q^* \in O_n^\alpha$ by $f - \phi Q = S_Q + X_n^p T_Q^*$ (where $S_Q \in \mathcal{F}_{n-1}[X_n]$ is of degree $< p$ in X_n and $T_Q^* \in \mathcal{F}_n$). It is immediate that $S_Q, T_Q^* \in O_n^\alpha$ and $u^{-1} T_Q^* \in O_n^\alpha$.

Define $T : O_n^\alpha \to O_n^\alpha$ by

$$T(Q) = u^{-1} T_Q^*.$$

Let $Q_1, Q_2 \in O_n^\alpha$. Then

$$||T(Q_1) - T(Q_2)|| < M \cdot ||T_{Q_1}^* - T_{Q_2}^*||$$

$$= \frac{M}{\alpha_n^p} \cdot ||X_n^p(T_{Q_1}^* - T_{Q_2}^*)||$$

$$\leq \frac{M}{\alpha_n^p} \cdot ||(S_{Q_1} - S_{Q_2}) + X_n^p(T_{Q_1}^* - T_{Q_2}^*)||$$

$$= \frac{M}{\alpha_n^p} \cdot ||\phi(Q_2 - Q_1)||$$

$$\leq \frac{M}{\alpha_n^p} \cdot ||\phi|| \cdot ||Q_2 - Q_1||$$

$$\leq \frac{1}{2} \cdot ||Q_2 - Q_1||$$

$$= \frac{1}{2} \cdot ||Q_1 - Q_2||.$$

Hence T is contractive and we finish the proof as in 5.6. $\qquad\square$

Corollary 7.7 (Weierstrass Preparation Theorem for O_n)   *Suppose $\Phi \in O_n$ is regular of order p (in X_n). Then there exists a (unique) unit $u \in O_n$ and*

$$f_0, \dots, f_{p-1} \in O_{n-1}$$

with $f_i(0) = 0$ such that

$$\Phi(X_1, \dots, X_n) = u(X_1, \dots, X_n) \cdot \left(X_n^p + \sum_{i=0}^{p-1} f_i(X_1, \dots, X_{n-1}) X_n^i \right).$$

(Of course we have the same equality for the corresponding functions.)

Proof   Same way as 5.7 is deduced from 5.6. $\qquad\square$

Corollary 7.8   *O_n is a Noetherian ring.*

Proof   By induction on n.

If $n = 0$, trivial. ($O_0 = K$ is a field.)

Suppose $n \geq 1$ and $I \trianglelefteq O_n$. Suppose I contains some element Φ which is regular in X_n of some order $p \in \mathbb{N}$. Let J be the ideal of O_n generated by Φ. Then $J \subseteq I$. Now every element of O_n, because of the division theorem, is equivalent (mod J) to an element of the form $\sum_{i=0}^{p-1} f_i X_n^i$ where $f_i \in O_{n-1}$. Hence O_n/J is generated as an O_{n-1}-module by $1, X_n, \dots, X_n^{p-1}$ and hence is Noetherian since O_{n-1} is Noetherian by the induction hypothesis. (See the remarks at the beginning of Section 2.) Hence the sub-O_{n-1}-module I/J of O_n/J is finitely generated by, say, $\Theta_1/J, \dots, \Theta_m/J$, $(\Theta_1, \dots, \Theta_m \in I)$. But then, every element of I is of the form $F_1 \Theta_1 + \cdots + F_m \Theta_m + G\Phi$ for some $F_1, \dots, F_m \in O_{n-1}(\subseteq O_n)$ and $G \in O_n$. So $\Theta_1, \dots, \Theta_m, G$ generate I as an O_n-module (i.e. as an ideal of O_n).

Now, in general, if $I \neq 0$ let $\Phi \in I$ be non-zero. By 6.9 there are $c_1, \dots, c_{n-1} \in K$ such that the map

$$\tau : f(X_1, \dots, X_n) \mapsto f(X_1 + c_1 X_n, \dots, X_{n-1} + c_{n-1} X_n, X_n)$$

is a homomorphism from $\mathcal{F}_n$ to $\mathcal{F}_n$ mapping Φ onto an element regular of some order p (in X_n). This map has an inverse

$$f(X_1, \dots, X_n) \mapsto f(X_1 - c_1 X_n, \dots, X_{n-1} - c_{n-1} X_n, X_n)$$

and is therefore a ring automorphism of $\mathcal{F}_n$. By 7.3 it restricts to an automorphism of O_n, whence the result by the first case. $\qquad\square$

We now turn to the Denef-van den Dries Preparation Theorem for O_n. The proof requires the following deep algebraic result, the proof of which is postponed to the next section.

Proposition 7.9 *Let $f_1, \ldots, f_s, f \in O_n \subseteq \mathcal{F}_n$ and suppose that the linear equation*

$$f_1 y_1 + \cdots + f_s y_s = f$$

is solvable in $\mathcal{F}_n$. Then it is solvable in O_n.

Proof Next section. □

Theorem 7.10 (Denef–van den Dries Preparation Theorem for O_n) *Let*

$$\Phi(X_1, \ldots, X_{m+n}) \in O_{m+n}.$$

Then there exists a positive integer d, elements $a_\nu(X_1, \ldots, X_m) \in O_m$ and units $u_\nu(X_1, \ldots, X_{m+n}) \in O_{m+n}$ for each $\nu \subset \mathbb{N}^n$ with $|\nu| < d$, such that

$$(\dagger) \qquad \Phi(X_1, \ldots, X_{m+n}) = \sum_{\nu \in \mathbb{N}^n, |\nu| < d} a_\nu(X_1, \ldots, X_m) X_*^\nu u_\nu(X_1, \ldots, X_{m+n})$$

where $X_* = (X_{m+1}, \ldots, X_{m+n})$.

Proof Let $R = O_m$ in 6.6 (allowed by 7.8). Then by 6.6 we can solve ($\dagger$) with a_ν's in O_m but with u_ν's in $O_m[[X_{m+1}, \ldots, X_{m+n}]] (\subseteq \mathcal{F}_{m+n})$.

We may write

$$(*) \qquad u_\nu(X_1, \ldots, X_{m+n}) = \beta_\nu + X_1 u_\nu^{(1)} + \cdots + X_{m+n} u_\nu^{(m+n)}$$

with the $u_\nu^{(i)}$'s in $\mathcal{F}_{m+n}$ and β_ν's in $K \setminus \{0\}$ (cf. 6.2). Substituting ($*$) into ($\dagger$) for each ν with $|\nu| < d$, we arrive at an equation of the form in 7.9 where we regard the $u_\nu^{(i)}$'s as 'variables'. But we can (and have already managed to) solve this equation in $\mathcal{F}_{m+n}$. Hence by 7.9 we can solve it in O_{m+n}. Now define the required units in O_{m+n} by the equation ($*$) (for $u_\nu^{(i)}$ the new convergent solutions). □

8 More on Adically Normed Rings and Modules

The purpose of this section is to develop further the theory of Noetherian rings up to the point that Proposition 7.9 may be easily deduced.

Let R be a ring and J any ideal of R. For $a \in R$ define

$$\text{ord}_J(a) = \begin{cases} \text{the largest } m \text{ such that } a \in J^m & \text{if such exists,} \\ \infty & \text{otherwise.} \end{cases}$$

Set $\|a\|_J = 2^{-\text{ord}_J(a)}$ ($= 0$ if $\text{ord}_J(a) = \infty$). Then it is easy to show that for all $a, b \in R$, $0 \le \|a\|_J \le 1$, $\|a+b\|_J \le \max(\|a\|_J, \|b\|_J)$ and $\|ab\|_J \le \|a\|_J\|b\|_J$. Hence $\| \cdot \|_J$ is a norm on R provided $a = 0$ whenever $\|a\|_J = 0$, i.e. provided that $\bigcap_{m \in \mathbb{N}} J^m = \{0\}$. Now one can easily derive a necessary condition for this to occur. For if $\| \cdot \|_J$ *is* a norm, then whenever $\|a\|_J < 1$ (i.e. $a \in J$) it cannot be the case that $1 + a$ is a zero divisor in R. For if $b(1 + a) = 0$, then $\|b\|_J = \| - b\, a\|_J \le \| - b\|_J\, \|a\|_J < \|b\|_J$. For R Noetherian this turns out to be sufficient, as we shall show in 8.4 below.

Lemma 8.1 (Artin-Rees Lemma)   *Let R be a Noetherian ring and M a finitely generated R-module. Suppose that N, N' are submodules of M and J is an ideal of R. Then there exists a natural number r_0 such that for all $n > r_0$*

$$J^n N \cap N' = J^{n-r_0}\left(J^{r_0} N \cap N'\right).$$

Remark   For any ideal I and submodules N_0, N_1 of an R-module, we always have $I(N_0 \cap N_1) \subseteq I N_0 \cap I N_1$.

Proof   The $\supseteq$ containment follows immediately from the remark. For the converse, we first treat the case $M = R$, i.e. when N, N' are ideals of R. Let $a_1, \ldots, a_m$ generate J. For each n, set

$$S_n = \{f(x_1, \ldots, x_m) \in R[x_1, \ldots, x_m] : f \text{ homogeneous of degree } n,$$
$$f(a_1, \ldots, a_m) \in J^n \cap N'\}.$$

Let $S = \bigcup_{n=1}^{\infty} S_n$ and let $\tilde{S}$ be the ideal of $R[x_1, \ldots, x_m]$ generated by S. By the Hilbert Basis Theorem we may choose $f_1, \ldots, f_s \in S$ generating $\tilde{S}$. Say f_i is homogeneous of degree d_i, so that $f_i \in S_{d_i}$ (for $i = 1, \ldots, s$). Let $r_0 = \max\{d_1, \ldots, d_s\}$. Suppose $n > r_0$ and $a \in J^n N \cap N'$. Then certainly $a \in J^n$ so $a = f(a_1, \ldots, a_m)$ for some $f(x_1, \ldots, x_m) \in R[x_1, \ldots, x_m]$ homogeneous of degree n (easy exercise). Since $a \in J^n N \cap N'$ we have $f \in S_n \subseteq \tilde{S}$, so we have an identity

$$f(x_1, \ldots, x_m) = \sum_{i=1}^{s} f_i(x_1, \ldots, x_m)\, g_i(x_1, \ldots, x_m)$$

for some $g_1, \ldots, g_s \in R[x_1, \ldots, x_m]$. Since f is homogeneous of degree n we may set the coefficient of each monomial in g_i not of degree $n - d_i$ to zero (for $i = 1, \ldots, s$) and retain the identity. Hence we may suppose that g_i is homogeneous of degree $n - d_i$, and therefore $g_i(a_1, \ldots, a_m) \in J^{n-d_i}$. Thus we have

$$
\begin{aligned}
g_i(a_1, \ldots, a_m) \cdot f_i(a_1, \ldots, a_m) &\in J^{n-d_i}(J^{d_i} \cap N') \quad \text{(since } f_i \in S_{d_i}) \\
&\subseteq J^{n-d_i-(r_0-d_i)}(J^{r_0-d_i} J^{d_i} N \cap J^{r_0-d_i} N') \quad \text{(by remark)} \\
&\subseteq J^{n-r_0}(J^{r_0} N \cap N').
\end{aligned}
$$

Thus $a = f(a_1, \ldots, a_m) \in J^{n-r_0}(J^{r_0} N \cap N')$ as required.

Let M now be an arbitrary finitely generated R-module. Consider the additive abelian group $R \times M$ and define multiplication by

$$
\langle r_1, m_1 \rangle \circ \langle r_2, m_2 \rangle = \langle r_1 r_2, r_1 m_2 + r_2 m_1 \rangle.
$$

Then one easily checks that $R \oplus M := \langle R \times M, +, \circ, \langle 0, 0 \rangle, \langle 1, 0 \rangle \rangle$ is a commutative ring with 1. Further, identifying R with $\{\langle r, 0 \rangle : r \in R\}$ and M with $\{\langle 0, m \rangle : m \in M\}$ we see that $R \oplus M$ is finitely generated over R, so Noetherian (by 3.3), and any R-submodule of M is an ideal of $R \oplus M$. The general result for R, N, N' now easily follows by the above result for ideals of $R \oplus M$.   □

Corollary 8.2   *Suppose R is a Noetherian ring, M a finitely generated R-module, J an ideal of R. Set $N = \bigcap_{n=0}^{\infty} J^n M$. Then $J N = N$.*

Proof   By 8.1 choose r_0 so that (setting $n = r_0 + 1$ in 8.1) $J^{r_0+1} M \cap N = J(J^{r_0} \cap N)$. Since $N \subseteq J^{r_0+1} M$, this gives $N = J N$.   □

Exercise 8.3   *Let A be an $s \times s$ matrix over a ring R. Then it is well known that A is an invertible matrix if and only if $\det(A)$ is a unit in R. The proof shows the following: suppose further that M is an R-module, $t_1, \ldots, t_s \in M$, and*

$$
A \begin{pmatrix} t_1 \\ \vdots \\ t_s \end{pmatrix} = \begin{pmatrix} 0 \\ \vdots \\ 0 \end{pmatrix}.
$$

Then $\det(A) \, t_i = 0$ for $i = 1, \ldots, s$.

Theorem 8.4 (Krull Intersection Theorem)   *Suppose R is a Noetherian ring, M a finitely generated R-module, J an ideal of R. Then $\bigcap_{n=0}^{\infty} J^n M = \{0\}$ if and only if for all $a \in J$ and for all $m \in M \setminus \{0\}$, $(1 + a)m \neq 0$.*

Proof   ($\Rightarrow$) If $a \in J$, $m \in M$ and $(1 + a)m = 0$ then for all odd n we have

$$(1 + a^n)m = (1 - a + a^2 - \cdots + a^{n-1})(1 + a)m = 0.$$

Hence $m \in J^n M$ for all n. Therefore $m = 0$.

($\Leftarrow$) Let $N = \bigcap_{n=0}^{\infty} J^n M$ and let $t_1, \ldots, t_s$ generate N (note: M is Noetherian – see section 3). Since $JN = N$ by the above corollary, we have $t_1, \ldots, t_s \in JN$ so

$$t_1 = a_{11}t_1 + \cdots + a_{1s}t_s$$

$$\vdots$$

$$t_s = a_{s1}t_1 + \cdots + a_{ss}t_s$$

for some $a_{ij} \in J$. Hence

$$\begin{pmatrix} a_{11} - 1 & a_{12} & \cdots & a_{1s} \\ a_{21} & a_{22} - 1 & \cdots & a_{2s} \\ \vdots & \vdots & \ddots & \vdots \\ a_{s1} & a_{s2} & \cdots & a_{ss} - 1 \end{pmatrix} \begin{pmatrix} t_1 \\ t_2 \\ \vdots \\ t_s \end{pmatrix} = 0.$$

But the determinant of this matrix clearly has the form $\pm(1 + a)$ for some $a \in J$. By 8.3, $(1 + a)t_i = 0$ for $i, \ldots, s$. Hence $t_i = 0$ by hypothesis, and so $N = 0$. $\square$

Corollary 8.5   *Let R be a Noetherian ring and I, J ideals of R not equal to R. Suppose that $1 + a$ is a unit in R for all $a \in J$. Then $\bigcap_{n=0}^{\infty}(I + J^n) = I$.*

Proof   Let $R' = R/I$. Then R' is a Noetherian ring. Let $h : R \to R'$ be the natural homomorphism and $J' = h[J]$. Then J' is an ideal of R' and if $a' \in J'$, say $h(a) = a'$ for some $a \in J$, then $(1 + a)b = 1$ for some $b \in R$ by the hypothesis. So $(1+a')\, h(b) = 1$, so $1+a'$ is a unit in R'. Hence, applying 8.4 we have $\bigcap_{n=0}^{\infty} J'^n = \{0\}$. But $h[I + J^n] \subseteq J'^n$ for all n, hence $h\left[\bigcap_{n=0}^{\infty}(I + J^n)\right] = \{0\}$. So $\bigcap_{n=0}^{\infty}(I + J^n) \subseteq \ker(h) = I$. The opposite inclusion is immediate. $\square$

Theorem 8.6   *Let $n \geq 1$. Let I be an ideal of $\mathcal{O}_n$ and let $\hat{I}$ be the ideal of $\mathcal{F}_n$ generated by I. Then $\hat{I} \cap \mathcal{O}_n = I$.*

Proof   The $\supseteq$ inclusion is obvious. For $\subseteq$ let J be the maximal ideal of $\mathcal{O}_n$, i.e. the ideal generated by $x_1, \ldots, x_n$. Suppose $g_1, \ldots, g_s$ generates I. Let $f \in \hat{I} \cap \mathcal{O}_n$. Then $f = f_1 g_1 + \cdots + f_s g_s$ for some $f_1, \ldots, f_s \in \mathcal{F}_n$. Let $r \in \mathbb{N}$ and write f_i as $p_i + h_i$, where p_i is a polynomial in $x_1, \ldots, x_n$ of degree less than r and $h_i \in \hat{J}^r$. Then $f = \sum_{i=1}^{n} p_i g_i + H$ where $H \in \hat{J}^r$. But $f - \sum_{i=1}^{n} p_i g_i \in \mathcal{O}_n$, so $H \in \mathcal{O}_n$. It is easy to show that the theorem holds for $\hat{J}^r$, i.e. $\mathcal{O}_n \cap \hat{J}^r = J^r$ (and $\hat{J^r} = \hat{J}^r$), hence $H \in J^r$. Thus $f \in I + J^r$. Since this holds for all $r \in \mathbb{N}$, we have, by 8.5 (and 7.4), $f \in I$ as required. $\qquad\qquad\square$

Proposition 7.9 is now immediate by applying the previous theorem to the ideal of $\mathcal{O}_n$ generated by $f_1, \ldots, f_s$.

9 The Denef–van den Dries Paper

For $r \in \mathbb{R}$, $r > 0$, $n \in \mathbb{N}$, $n > 0$, let B_r^n (respectively $\overline{B_r^n}$) denote the set

$$\{\langle x_1, \ldots, x_n \rangle \in \mathbb{R} : |x_i| < r \text{ (respectively } \leq) \text{ for } i = 1, \ldots, n\}.$$

The language L_{an} consists of $+, \cdot, -, <$ together with a constant symbol for each real number and a function symbol for each $n \in \mathbb{N}$, $n > 0$, $r \in \mathbb{R}$, $r > 0$ and $f \in \mathcal{O}_n$ with $\overline{B_r^n} \subseteq \mathrm{dom}(f)$. The structure $\mathbb{R}_{an}$ interprets the former symbols naturally and the function symbols of the latter kind as the functions $\mathbb{R}^n \to \mathbb{R}$,

$$\bar{x} \mapsto \begin{cases} \widetilde{f}(\bar{x}) & \text{if } \bar{x} \in B_r^n \\ 0 & \text{otherwise} \end{cases}.$$

(Recall that $\widetilde{f}$ denotes the function determined by the convergent power series f.)

L_{an}^D denotes the language L_{an} together with an additional binary function symbol D, and then $\mathbb{R}_{an}^D$ denotes the expansion of $\mathbb{R}_{an}$ obtained by interpreting D as

$$\langle x, y \rangle \mapsto \begin{cases} \frac{x}{y} & \text{if } y \neq 0 \\ 0 & \text{otherwise} \end{cases}.$$

Clearly $\mathbb{R}_{an}^D$ and $\mathbb{R}_{an}$ have the same definable sets. Let T_{an}^D and T_{an} denote the complete theories of these structures.

Theorem 9.1 *T_{an} is model complete; T_{an}^D eliminates quantifiers.*

Exercise 9.2 *Deduce the first statement of the previous theorem from the second one.*

Example 9.3 (Osgood 1910) *T_{an} does not eliminate quantifiers.*

Let

$$e(x) = \begin{cases} e^x & \text{for } |x| \le 1, \\ 0 & \text{otherwise.} \end{cases}$$

Let $\varphi(a,b,c) \Leftrightarrow \exists x \exists y (a = x \wedge b = xy \wedge c = x\,e(y) \wedge |x|,|y| \le 1)$. Then φ is not equivalent to any quantifier-free L_{an}-formula. For if it were then it is not too hard to see (using a little analysis) that there would have to be an $f \in \mathcal{O}_3$ such that for all x,y sufficiently close to 0 we have $\tilde{f}(x,xy,x\,e(y)) = 0$, but $f \not\equiv 0$. Writing $f(x_1,x_2,x_3) = \sum_{i=0}^{\infty} h_i(x_1,x_2,x_3)$ where h_i is homogeneous of degree i we have

$$0 \equiv \sum_{i=0}^{\infty} h_i(x, xy, x\,e(y)) = \sum_{i=0}^{\infty} x^i\, h_i(1, y, e(y))$$

close to 0. Thus for all y, the series in x is 0. Hence $h_i(1, y, e(y)) = 0$ for all i and y, with y close to 0. But y and e^y are algebraically independent functions (over $\mathbb{R}$) – exercise. Hence $h_i(1, x_2, x_3) \equiv 0$, from which it follows that $h_i(x_1, x_2, x_3) \equiv 0$ and so $f \equiv 0$, a contradiction.

Of course, $\varphi(a,b,c)$ is equivalent to the quantifier-free L_{an}^D-formula

$$(a = b = c = 0) \vee (a \ne 0 \wedge D(c,a) = e(D(b,a)) \wedge |a| \le 1 \wedge |D(b,a)| \le 1).$$

Remark (Original work in the 60s by Gabrielov and Łojasiewicz) Łojasiewicz extended the notion of semi-algebraic set[1] to the analytic category. The right definition is as follows: A subset $X \subseteq \mathbb{R}^n$ is called semi-analytic if for all $\bar{a} \in \mathbb{R}^n$ there exists an open neighborhood $U_{\bar{a}}$ of $\bar{a}$ in $\mathbb{R}^n$ such that $X \cap U_{\bar{a}}$ can be expressed as a Boolean combination of sets of the form

$$\{\bar{x} \in U_{\bar{a}} : \tilde{f}(\bar{a} - \bar{x}) > 0\}$$

for $f \in \mathcal{O}_n$ with $\bar{a} - U_{\bar{a}} \subseteq \text{dom}(f)$.

Exercise 9.4 *A subset $X \subseteq \mathbb{R}^n$ is semi-analytic if and only if $\overline{B_r^n} \cap X$ is quantifier-free definable in $\mathbb{R}_{an}$ for all $r \in \mathbb{R}$.*

One wanted to show that the semi-analytic sets had good properties (similar to the class of semi-algebraic sets). Unfortunately, 9.3 shows that the class of semi-analytic sets is not closed under projections. Gabrielov then defined a subset $X \subseteq \mathbb{R}^n$ to be sub-analytic if for all $\bar{a} \in \mathbb{R}^n$ there exists a neighborhood $U_{\bar{a}}$ of a, $m \in \mathbb{N}$ and a bounded semi-analytic $Y \subseteq \mathbb{R}^{n+m}$ such that $\pi[Y] = U_{\bar{a}} \cap X$. He showed that the complement of a sub-analytic set is sub-analytic.

---

[1] i.e. quantifier-free definable set in $\overline{\mathbb{R}} = \langle \mathbb{R}; 0, 1, +, \cdot, -, < \rangle$.

Exercise 9.5 *Using 9.1 show (from the definition above) that a subset $X \subseteq \mathbb{R}^n$ is sub-analytic if and only if $\overline{B_r^n} \cap X$ is quantifier-free definable in $\mathbb{R}_{an}^D$ for all $r \in \mathbb{R}$. Deduce Gabrielov's theorem.*

I also mention here the following important result. One can use it to prove that all $\mathbb{R}_{an}^D$-definable unary functions have Puiseax expansions at ∞.

Theorem 9.6 (Łojasiewicz) *A sub-analytic subset of $\mathbb{R}^2$ is semi-analytic, i.e. (by above) any $\mathbb{R}_{an}$ (or $\mathbb{R}_{an}^D$)-definable subset of $\mathbb{R}^2$ is quantifier-free $\mathbb{R}_{an}$-definable.*

9.1 Definable subsets of $\mathbb{R}$ in $\mathbb{R}_{an}$

Lemma 9.7 *For all $f \in \mathcal{O}_1 \setminus \{0\}$ there exist $r \in \mathbb{N}$, $g \in \mathcal{O}_1$, g a unit with $\mathrm{dom}(g) = \mathrm{dom}(f)$, such that for all $x \in \mathrm{dom}(g)$, $\tilde{f}(x) = x^r \tilde{g}(x)$. Hence $\tilde{f}$ has constant non-zero sign on $(0, \varepsilon)$ for some $\varepsilon > 0$.*

Proof Obvious.   □

Lemma 9.8 *For all terms $t(x)$ of L_{an}^D there exists $\varepsilon > 0$ such that either*

1. $t(x) = 0$ *for all $x \in (0, \varepsilon)$, or*
2. $t(x) = x^n \tilde{f}(x)$ *for all $x \in (0, \varepsilon)$, for some unique (possibly negative) $n \in \mathbb{Z}$ and unit $f \in \mathcal{O}_1$ with $(-\varepsilon, \varepsilon) \subseteq \mathrm{dom}(f)$.*

(We also use t to denote the function it defines in $\mathbb{R}_{an}^D$.)

Proof This is obvious if $t(x) = x$ or a constant. Also, if it is true for some $t_1(x), t_2(x)$ then it is true for $t_1(x) \pm t_2(x)$ and $t_1(x) \cdot t_2(x)$ (using 9.7). Also, if (1) holds for t_1 or t_2 then (1) holds for $D(t_1, t_2)$. If (2) holds for t_1 and t_2 we may choose (using 9.7) some ε small enough so that $t_2(x) \neq 0$ on $(0, \varepsilon)$. It easily follows that (2) holds for $D(t_1, t_2)$ (since the ratio of units is a unit).

Now suppose that the result holds for $t_1, \ldots, t_n$ and $f \in \mathcal{O}_n$, $\alpha \in \mathbb{R}$, $\alpha > 0$ and $\overline{B_\alpha^n} \subseteq \mathrm{dom}(f)$. We must consider

$$\begin{cases} \tilde{f}(t_1(x), \ldots, t_n(x)) & \text{if } |t_i(x)| < \alpha \text{ for all } i = 1, \ldots, n \\ 0 & \text{otherwise} \end{cases}$$

Now if (1) holds for some t_i we can reduce n. So we can choose ε small enough, units $f_1, \ldots, f_n \in \mathcal{O}_1$ and $m_1, \ldots, m_n \in \mathbb{Z}$ such that $t_i(x) = x^{m_i} \tilde{f}_i(x)$ for all $x \in (0, \varepsilon)$. If $m_i < 0$ for some i then (since $\tilde{f}_i(x)$ is bounded away from 0) we may suppose $|t_i(x)| > \alpha$ for all $x \in (0, \varepsilon)$ so (1) holds for the composite term. Hence we may suppose that $m_i \geq 0$ for $i = 1, \ldots, n$. Now by 9.7 $\alpha + t_i(x)$ and $t_i(x) - \alpha$ have constant sign on $(0, \varepsilon)$. If the former is nonpositive or the latter nonnegative then (1) holds for the composite term. Otherwise $-\alpha < t_i(x) < \alpha$ for all $x \in (0, \varepsilon)$ and for $i = 1, \ldots, n$. Let $a_i = (x^{m_i} \widetilde{f_i}(x))_{x=0}$. Then $-\alpha \leq a_i \leq \alpha$. So $\langle a_1, \ldots, a_n \rangle \in \overline{B_\alpha^n} \subseteq \mathrm{dom}(f)$. So it follows that the composite term is equal to $\widetilde{g}(x)$ for all $x \in (0, \varepsilon)$ for some $g \in \mathcal{O}_1$. Hence (2) holds for the composite term with $m \geq 0$ (using 9.7). $\qquad\square$

Corollary 9.9 *Let $t(x)$ be a term of L_{an}^D. Then there exists a partition of $\mathbb{R}$ into finitely many open intervals and points such that $t(x)$ has constant sign on each such interval.*

Proof By 9.8 (and 9.7), if $a \in \mathbb{R}$ then there exists $\varepsilon_a > 0$ such that $t(x)$ has constant sign on $(a, a + \varepsilon_a)$ and $(a - \varepsilon_a, a)$ (consider the terms $t(a + x)$ and $t(a - x)$). Also there exists $\eta > 0$ such that $t(x)$ has constant sign on $(1/\eta, \infty)$ and $(-\infty, -1/\eta)$ (consider the terms $t(D(1, x))$ and $t(-D(1, x))$). The result follows by the compactness of the closed interval $[-1/\eta, 1/\eta]$. $\qquad\square$

Theorem 9.10 (Assuming 9.1) *Every subset of $\mathbb{R}$ definable in $\mathbb{R}_{an}^D$ is a finite union of open intervals and points. So T_{an}^D is o-minimal.*

Proof Immediate from 9.1 and 9.9. $\qquad\square$

9.2 The proof that T_{an}^D eliminates quantifiers

Let $M_1, M_2 \models T_{an}^D$. Suppose that K is a $(L_{an}^D$-) substructure of M_1, that $e : K \to M_2$ is an embedding, and that M_2 is sufficiently saturated. By 1.5. we must consider $a \in M_1$ and extend e to $e' : K\langle a \rangle \to M_2$ where $K\langle a \rangle$ denotes the L_{an}^D-substructure of M_1 generated by K and a.

Note that we may suppose $\mathbb{R}_{an}^D \subseteq_{L_{an}^D} K$ (since we have all $r \in \mathbb{R}$ as a constant symbol of L_{an}^D), and we may suppose e is the identity on $\mathbb{R}_{an}^D$.

Let

$$\mu = \{\alpha \in M_1 : |\alpha| < r \text{ for all } r \in \mathbb{R}, r > 0\}.$$

Now for $n \in \mathbb{N}, f \in \mathcal{O}_n$, and $r \in \mathbb{R}_{>0}$, with $\overline{B_r^n} \subseteq \mathrm{dom}(f)$ (which we abbreviate by saying that (n, r, f) is *acceptable*), let us denote by f_r the interpretation of the corresponding function symbol in M_1 and M_2. Certainly for any such suitable r, f_r defines a natural, untruncated function $\mu^n \to M_1$.

Lemma 9.11 *With the above notation, suppose that there exists a map* e'' :
$K\langle a\rangle \cap \mu \to M_2$ *extending* $e \upharpoonright K \cap \mu$ *such that for all* $n \in \mathbb{N}$, *for all* $\alpha_1, \ldots, \alpha_n \in$
$K\langle a\rangle \cap \mu$ *and for all* f, r *as above,*

$$M_1 \models f_r(\alpha_1, \ldots, \alpha_n) > 0 \Rightarrow M_2 \models f_r(e''(\alpha_1), \ldots, e''(\alpha_n)) > 0$$
$$M_1 \models f_r(\alpha_1, \ldots, \alpha_n) = 0 \Rightarrow M_2 \models f_r(e''(\alpha_1), \ldots, e''(\alpha_n)) = 0.$$

Then e'' *extends to an* L_{an}^D-*embedding* $e' : K\langle a\rangle \to M_2$ *(which also extends* e*).*

Proof If $b \in K\langle a\rangle$, then either $s + b \in \mu$ for some (unique) $s \in \mathbb{R}$, and we
set $e'(b) = e''(s + b) - s$; or b is infinite and $1/b \in \mu$, in which case we set
$e'(b) = 1/e''(1/b)$. It is easy to check that e' is an ordered field embedding
$K\langle a\rangle \to M_2$ extending e. (Note that the $\mathcal{O}_n$'s include all polynomials.) In
particular e' preserves D.

Suppose f, r as above. Let $\alpha_1, \ldots, \alpha_n \in K\langle a\rangle$. If $|\alpha_i| \geq r$ for some i, then
$f_r(\alpha_1, \ldots, \alpha_n) = 0$. But $|e'(\alpha_i)| \geq e'(r) = r$ (since e' is an ordered field
embedding extending e), therefore $f_r(e'(\alpha_1), \ldots, e'(\alpha_n)) = 0$. If $|\alpha_i| < r$ for
$i = 1, \ldots, n$, let $\alpha_{n+1} = f_r(\alpha_1, \ldots, \alpha_n)$ and write $\alpha_i = \beta_i - s_i$ for $i = 1, \ldots, n+$
1, where $\beta_i \in \mu$, $s_i \in \mathbb{R}$. Note that $|s_i| \leq r$ for $i = 1, \ldots, n$, and β_{n+1}, s_{n+1}
exists because f_r is bounded and hence α_{n+1} is finite.

Define $k : \mathbb{R}^{n+1} \to \mathbb{R}$ by $k(x_1, \ldots, x_{n+1}) = \tilde{f}(x_1 - s_1, \ldots, x_n - s_n) -$
$x_{n+1} + s_{n+1}$. Then $k = \tilde{g}$ for some $g \in \mathcal{O}_{n+1}$ since $\overline{B}_r^n \subseteq \mathrm{dom}(f)$ and so
$\langle s_1, \ldots, s_n \rangle \in \mathrm{dom}(f)$. Now choose $\varepsilon \in \mathbb{R}$, $\varepsilon > 0$, such that $\overline{B}_\varepsilon^{n+1} \subseteq \mathrm{dom}(g)$ and
$\overline{B}_{r+\varepsilon}^n \subseteq \mathrm{dom}(f)$. Then $0 = g_\varepsilon(\beta_1, \ldots, \beta_{n+1})$, so $0 = g_\varepsilon(e''(\beta_1), \ldots, e''(\beta_{n+1}))$
(by hypothesis). But

$$T_{an}^D \models \forall x_1, \ldots, x_{n+1}(\max |x_i| < \varepsilon \to g_\varepsilon(x_1, \ldots, x_{n+1})$$
$$= f_{r+\varepsilon}(x_1 - s_1, \ldots, x_n - s_n) - x_{n+1} + s_{n+1}).$$

Hence

$$0 = f_{r+\varepsilon}(e''(\beta_1) - s_1, \ldots, e''(\beta_n) - s_n) - e''(\beta_{n+1}) + s_{n+1}$$
$$= f_{r+\varepsilon}(e'(\alpha_1), \ldots, e'(\alpha_n)) - e'(\alpha_{n+1})$$
$$= f_r(e'(\alpha_1), \ldots, e'(\alpha_n)) - e'(\alpha_{n+1}).$$

(The second equality follows from the definition of e', while the third holds
since $|e'(\alpha_i)| < r$ for $i = 1, \ldots, n$ as e' is an ordered field embedding.)
Hence in all cases, $e'(f_r(\alpha_1, \ldots \alpha_n)) = f_r(e'(\alpha_1), \ldots, e'(\alpha_n))$. So e' is an
L_{an}^D-embedding as required. $\qquad\square$

Exercise 9.12   *Suppose $M_1, M_2 \models T_{an}^D$ with $K \subseteq M_1$, $e : K \to M_2$ as above. Then for all acceptable $(m + n, r, f)$ and $c_1, \ldots, c_m \in K \cap \mu$ we have*

$$M_1 \models \forall \bar{y} \in \mu\, f_r(\bar{c}, \bar{y}) = 0 \Rightarrow M_1 \models \forall \bar{y} f_r(\bar{c}, \bar{y}) = 0$$
$$\Rightarrow M_2 \models \forall \bar{y} f_r(e(\bar{c}), \bar{y}) = 0.$$

To prove 9.1 it is clearly sufficient, in view of 9.11 and the saturation of M_2, to establish the following:

Lemma 9.13   *Let $\varphi(\bar{x}, \bar{y}) = \varphi(x_1, \ldots, x_m, y_1, \ldots, y_n)$ be a conjunction of formulas of the form $f_r(\bar{x}, \bar{y}) > 0$ or $f_r(\bar{x}, \bar{y}) = 0$, where $(m + n, r, f)$ is acceptable. Suppose $c_1, \ldots, c_m \in \mu \cap K$, $\alpha_1, \ldots, \alpha_n \in \mu$, and $M_1 \models \varphi[\bar{c}, \bar{\alpha}]$. Then $\exists \beta_1, \ldots, \beta_n \in M_2$ such that $M_2 \models \varphi[e(\bar{c}), \bar{\beta}]$.*

Proof   By induction on n. The case $n = 0$ is trivial, but we do the case $n = 1$. We may clearly suppose that all those r's for which f_r occurs in φ are the same, say r.

Let S be the set of $f \in \mathcal{O}_{n+1}$ such that f_r occurs in φ. Let $c_1, \ldots, c_m \in \mu \cap K$, $\alpha_1 \in \mu$ be such that $M_1 \models \varphi[c_1, \ldots, c_n, \alpha_1]$. Let $f \in S$.

By 7.10, there exist $\alpha \in \mathbb{N}$, $d > 0$, $a_i(\bar{x}) \in \mathcal{O}_m$, and units $u_i(\bar{x}, y_1) \in \mathcal{O}_{m+1}$ ($\forall i < d$) such that in $\mathcal{O}_{m+1}$ we have the identity:

$$f(\bar{x}, y_1) = \sum_{i < d} a_i(\bar{x}) y_1^i u_i(\bar{x}, y_1).$$

Clearly we may suppose that it is the same d for all $f \in S$.

Let $\varepsilon \in \mathbb{R}$, $\varepsilon > 0$ be small enough so that all $F \in \mathcal{O}_N$ considered in this proof are such that (N, ε, F) is acceptable.

Now work in M_1. We have for all $y_1 \in M_1$, $|y_1| < \varepsilon$ implies

$$f_\varepsilon(\bar{c}, y_1) = \sum_{i < d} a_{i,\varepsilon}(\bar{c}) y_1^i u_{i,\varepsilon}(\bar{c}, y_1). \tag{6.1}$$

If $a_i(\bar{c}) = 0$ for all $i < d$, then also $a_i(e(\bar{c})) = 0$ for all $i < d$ and we may omit all atomic formulas involving f from φ.

Otherwise, choose $i_0 < d$ such that $0 \neq |a_{i_0,\varepsilon}(\bar{c})| \geq |a_{i,\varepsilon}(\bar{c})|$ for all $i < d$. Let

$$k_i' = \frac{a_{i,\varepsilon}(\bar{c})}{|a_{i_0,\varepsilon}(\bar{c})|} \tag{6.2}$$

for all $i < d$. Since K is a field, $k_i' \in K$ for $i < d$. Furthermore, $|k_i'| \leq 1$ for all $i < d$ and $k_{i_0}' = \pm 1$. Define k_i's by:

$$k_i' = k_i + s_i \text{ where } k_i \in \mu \cap K \text{ and } s_i \in \mathbb{R} \text{ for all } i < d. \tag{6.3}$$

Note that $k_{i_0} = 0$ and $s_{i_0} = \pm 1$.

Define $g \in \mathcal{O}_{m+1+(d-1)}$, with $\bar{z} = \{z_i \mid i < d, i \neq i_0\}$ the new variables, by

$$g(\bar{x}, y_1, \bar{z}) = s_{i_0} u_{i_0}(\bar{x}, y_1) y_1^{i_0} + \sum_{i<d, i \neq i_0} u_i(\bar{x}, y_1)(z_i + s_i) y_1^i. \quad (6.4)$$

Then for all $y_1 \in M_1$, $|y_1| < \varepsilon$, we have (from (2)–(5))

$$f_\varepsilon(\bar{c}, y_1) = |a_{i_0, \varepsilon}(\bar{c})| g(\bar{c}, y_1, \bar{k}). \quad (6.5)$$

Now choose $p \in \mathbb{N}$ minimal such that $s_p \neq 0$. Obviously p exists and $p \leq i_0$. Now clearly g is regular in y_1 of order p, hence (by 7.7) (in $\mathcal{O}_{m+1+(d-1)}$):

$$g(\bar{x}, y_1, \bar{z}) = (y_1^p + h_1(\bar{x}, \bar{z}) y_1^{p-1} + \ldots + h_p(\bar{x}, \bar{z})) Q(\bar{x}, y_1, \bar{z}) \quad (6.6)$$

for some $h_i \in \mathcal{O}_{m+d-1}$ and a unit $Q \in \mathcal{O}_{m+1+(d-1)}$.
By (6.5), for all $y_1 \in M_1$, $|y_1| < \varepsilon$ implies

$$f_\varepsilon(\bar{c}, y_1) = |a_{i_0, \varepsilon}(\bar{c})| Q_\varepsilon(\bar{c}, y_1, \bar{k})(y_1^p + \ldots + h_p(\bar{c}, \bar{k})). \quad (6.7)$$

Now we may suppose ε has been chosen small enough so that

$$T_{\mathrm{an}}^D \vdash \forall \bar{x}, \bar{z}, y_1 \in \bar{B}_\varepsilon \left(|Q_\varepsilon(\bar{x}, y_1, \bar{z}) - Q_\varepsilon(\bar{0}, 0, \bar{0})| < \frac{1}{2} |Q_\varepsilon(\bar{0}, 0, \bar{0})| \right). \quad (6.8)$$

By the exercise, (6.7) also holds in M_2 upon applying e to the parameters–note they all lie in $\mu \cap K$.

By (6.7), (6.8), $\varphi(\bar{c}, y_1)$ is equivalent in M_1 to some $\psi(\overline{h_\varepsilon(\bar{c}, \bar{k})}, y_1)$ for some (quantifier-free) formula $\psi(t_1, \ldots, t_q, y_1)$ of $\bar{L}$ (where $q = \sum_{f \in \mathcal{S}} p$) and $\varphi(e(\bar{c}), y_1)$ is equivalent in M_2 to $\psi(\overline{h_\varepsilon(e(\bar{c}), e(\bar{k}))}, y_1)$, for all $y_1 \in M_2$ with $|y_1| < \varepsilon$.

Now $M_1 \models \exists y_1 (|y_1| < \varepsilon \wedge \varphi(\bar{c}, y_1))$ (namely $y_1 = \alpha_1$), so

$$M_1 \models \exists y_1 \left(|y_1| < \varepsilon \wedge \psi \left(\overline{h_\varepsilon(\bar{c}, \bar{k})}, y_1 \right) \right).$$

But $\exists y_1 (|y_1| < \varepsilon \wedge \psi(\bar{t}, y_1))$ is equivalent in M_1 and M_2 to a quantifier-free formula of $\bar{L}$ (since $M_1 \upharpoonright \bar{L}, M_2 \upharpoonright \bar{L} \equiv \mathbb{R}$), so:

$$M_2 \models \exists y_1 \left(|y_1| < \varepsilon \wedge \psi \left(e \left(\overline{h_\varepsilon(\bar{c}, \bar{k})} \right), y_1 \right) \right)$$

(since e is certainly an $\bar{L}$-embedding), and then

$$M_2 \models \exists y_1 \left(|y_1| < \varepsilon \wedge \psi \left(h_\varepsilon(e(\bar{c}), e(\bar{k})), y_1 \right) \right)$$

(since e is an L_{an}-embedding), and finally $M_2 \models \exists y_1 \varphi(e(\bar{c}), y_1)$ by the previous paragraph.

For the inductive step we proceed exactly as above to arrive at (6.4):

$$g(\bar{x}, \bar{y}, \bar{z}) = s_{v^0} u_{v^0}(\bar{x}, \bar{y}) y^{v^0} + \sum_{\substack{v \in \mathbb{N}^{n+1} \\ |v| < d \\ v \neq v^0}} u_v(\bar{x}, \bar{y})(z_v + s_v) \bar{y}^v$$

where now $\bar{y} = y_1, \ldots, y_{n+1}$.

We want to write g in the form (6.6), but it might not be regular in any of the $\bar{y}$-variables. To resolve this difficulty we define

$$\Lambda(\bar{y}) = \langle y_1 + y_{n+1}^{d^n}, y_2 + y_{n+1}^{d^{n-1}}, \ldots, y_n + y_{n+1}^{d}, y_{n+1} \rangle.$$

This is a bijection from $\mu^{n+1} \to \mu^{n+1}$ with inverse

$$\Omega(\bar{y}) = \langle y_1 - y_{n+1}^{d^n}, \ldots, y_n - y_{n+1}^{d}, y_{n+1} \rangle.$$

Now

$$g(\bar{0}, \Lambda(0, \ldots, 0, y_{n+1}), \bar{0}) = \sum_{\substack{v \in \mathbb{N}^{n+1} \\ |v| < d}} u_v(\bar{0}, \Lambda(\bar{0}, y_{n+1})) s_v y_{n+1}^{v_1 d^n + v_2 d^{n-1} + \ldots + v_n d + v_{n+1}}.$$

Since the exponents of y_{n+1} here are all distinct it follows that if v is lexicographically minimal such that $s_v \neq 0$ (v exists since $s_{v^0} \neq 0$) and $p = v_1 d^n + \ldots + v_{n+1}$ (where $v = \langle v_1, \ldots, v_{n+1} \rangle$). Then $g(\bar{x}, \Lambda(y_1, \ldots, y_{n+1}), \bar{z})$ is regular in y_{n+1} of order p.

Now, as above, we have (6.7) in the form

$$\forall y_1, \ldots, y_{n+1} \in M_1, |y_1|, \ldots, |y_{n+1}| < \varepsilon \Rightarrow$$

$$f_\varepsilon(\bar{c}, \Lambda(y_1, \ldots, y_{n+1}))$$
$$= |a_{v^0, \varepsilon}(\bar{c})| Q_\varepsilon(\bar{c}, y_1, \ldots, y_{n+1}, \bar{k})(y_{n+1}^p + \ldots + h_{p, \varepsilon}(\bar{c}, \bar{k}, y_1, \ldots, y_n)). \quad (6.9)$$

Notice that the transformation Λ does not depend on $f \in \mathcal{S}$.

Now, writing '$\exists_\varepsilon u \ldots$' for the quantifier '$\exists u(|u| < \varepsilon \wedge \ldots$', we have that $\exists_\varepsilon y_1, \ldots, y_{n+1} \varphi(\bar{c}, y_1, \ldots, y_{n+1})$ is equivalent to $\exists_\varepsilon y_1, \ldots, y_{n+1} \varphi(\bar{c}, \Lambda(y_1, \ldots, y_{n+1}))$, and the latter is equivalent to some $\exists_\varepsilon y_1, \ldots, y_{n+1} \psi(\bar{c}, \bar{k}, y_1, \ldots, y_{n+1})$ where y_{n+1} only occurs polynomially. This, by Tarski, is equivalent to $\exists_\varepsilon y_1, \ldots, y_n \psi(\bar{c}, \bar{k}, y_1, \ldots, y_n)$ for some quantifier-free formula ψ of L_{an}. Moreover, as above, these equivalences hold in M_2 with e applied to parameters (using exercise 9.12 on (6.9)). The result follows by induction on n. $\qquad\square$

7

Relative Manin-Mumford for abelian varieties

D. Masser

2010 MSC codes. 11G10, 14K15, 14K20, 11G50, 34M99.

Abstract

With an eye or two towards applications to Pell's equation and to Davenport's work on integration of algebraic functions, Umberto Zannier and I have recently characterized torsion points on a fixed algebraic curve in a fixed abelian scheme of dimension bigger than one (when all is defined over the algebraic numbers): there are at most finitely many points provided the natural obstacles are absent. I sketch the proof as well as the applications.

A very simple problem of Manin-Mumford type is: find all roots of unity λ, μ with $\lambda + \mu = 1$. Here the solution is easy: we have $|\lambda| = |1 - \lambda| = 1$ and so in the complex plane λ lies on the intersection of two circles; in fact λ must be one of the two primitive sixth roots of unity (the picture doesn't work too well in positive characteristic, and indeed any non-zero element of any finite field is already a root of unity, so from now on we stick to zero characteristic). This result has something to do with the multiplicative group $\mathbf{G_m}$, which can be regarded as $\mathbf{C}^*$. Actually with $\mathbf{G_m^2}$ and the "line" inside it parametrized by $P = (\lambda, 1 - \lambda)$: we ask just that P is torsion.

Now it is easy to generalize, at least the problem, to other algebraic varieties in other commutative algebraic groups.

O-Minimality and Diophantine Geometry, ed. G. O. Jones and A. J. Wilkie. Published by Cambridge University Press. © Cambridge University Press 2015.

For example let E be the elliptic curve defined by $y^2 = x(x-1)(x-4)$. Asking for all complex λ such that the points

$$(2\lambda, \sqrt{2\lambda(2\lambda-1)(2\lambda-4)}), \quad (3\lambda, \sqrt{3\lambda(3\lambda-1)(3\lambda-4)}) \qquad (7.1)$$

are both torsion amounts to asking for torsion points on a certain curve in the surface E^2. But here the solution is much more difficult (and it is not clear to me that one can find all λ explicitly as above).

It was Hindry [H] who solved the general problem with any algebraic variety in any commutative algebraic group G. The outcome for a curve in G is that it contains at most finitely many torsion points unless one of its components is a connected one-dimensional "torsion coset"; that is, a translate $P_0 + H$ of an algebraic subgroup H of G by a torsion point P_0. This H contains infinitely many torsion points and so $P_0 + H$ also.

Thus for $G = \mathbf{G}_{\mathrm{m}}^2$ the analogue for $\lambda\mu = 1$ of the problem above will not lead to finiteness, as the curve is such an H. Similarly $\lambda\mu = -1$ is $P_0 + H$ for $P_0 = (1, -1)$ with $2P_0 = 0$ (written additively).

More generally G can be $\mathbf{G}_{\mathrm{m}}^n$, E^n as in Habegger's talks in this volume, or an abelian variety A as in Orr's article, or products of these, or "twisted products" sitting inside an exact sequence

$$0 \to T \to G \to A \to 0 \qquad (7.2)$$

where T is a power of $\mathbf{G}_{\mathrm{m}}$ or even a product $\mathbf{G}_{\mathrm{a}}^r \times \mathbf{G}_{\mathrm{m}}^s$ for the additive group $\mathbf{G}_{\mathrm{a}} = \mathbf{C}$. Here the twisting can be quite complicated and G can end up very far from just $T \times A$. It is classical that every commutative algebraic group over $\mathbf{C}$ has this form. We will see how several types turn up in applications.

The applications involve most naturally the "relative case", where G itself is allowed to vary in a family. Most of the current results allow only a single parameter here, and we already have parameters in the algebraic variety, so this had better stay a curve, essentially with a parameter λ, and G had better depend on no more than λ. An example like (1) involves the points

$$(2, \sqrt{4-2\lambda}), \quad (3, \sqrt{18-6\lambda}) \qquad (7.3)$$

now on the elliptic curve E_λ defined by $y^2 = x(x-1)(x-\lambda)$; that is the famous Legendre family. Again we go to E_λ^2, where the square is the "fibre square" defined by the equations

$$y_1^2 = x_1(x_1-1)(x_1-\lambda), \quad y_2^2 = x_2(x_2-1)(x_2-\lambda)$$

with λ in common, and we get a curve defined by $x_1 = 2, x_2 = 3$. Then by [MZ1] there are at most finitely complex values of λ (now not 0, 1 so that we have a genuine elliptic curve) such that both points in (3) are torsion. Here their effective determination may be a difficult problem in practice and even in principle.

In various works Umberto Zannier and I have treated any curve in any parametrized abelian variety A_λ of "relative dimension" at least two, sometimes with the proviso that everything is defined over the field $\overline{\mathbf{Q}}$ of all algebraic numbers. We get finiteness with a similar condition about torsion cosets, now interpreted schemewise or more intuitively "identically in λ". For example in E_λ^2 above these are defined essentially by the vanishing of non-trivial integral linear combinations $n_1(x_1, y_1) + n_2(x_2, y_2)$. It is not difficult to show that this is impossible for (3). On the other hand

$$2(0,0) = 2(1,0) = 2(\lambda, 0) = 4(\sqrt{\lambda}, i(\lambda - \sqrt{\lambda})) = 0$$

so it is not quite easy.

A start has been made on more general G_λ. First with Bertrand and Pillay we have considered

$$1 \to \mathbf{G}_{\mathrm{m}} \to G_\lambda \to E_\lambda \to 0 \tag{7.4}$$

and

$$1 \to \mathbf{G}_{\mathrm{m}} \to G_\lambda \to E \to 0 \tag{7.5}$$

with E not depending on λ. Bertrand [Be] had already given a surprising counterexample in (5): in rather special situations it is possible to construct what he calls a "Ribet curve" having infinitely many torsion points, even though it is not a torsion coset. We then checked in [BMPZ] that this happens only for Ribet curves.

And Harry Schmidt in Basle [Sc] has done

$$0 \to \mathbf{G}_{\mathrm{a}} \to G_\lambda \to E_\lambda \to 0 \tag{7.6}$$

where it is reassuring to find that there are no counterexamples.

We give a short proof sketch of the result for (3). As in Habegger's talks, it hinges on the analytic representation of an elliptic curve as a quotient of $\mathbf{C}$ by a lattice, as in the general strategy of Zannier expounded in [PZ]. For E_λ this lattice Ω_λ depends on λ, and in fact one can take a basis of periods

$$\Omega_\lambda = \mathbf{Z} f_\lambda + \mathbf{Z} g_\lambda$$

where f_λ is hypergeometric $\pi F(\frac{1}{2}, \frac{1}{2}, 1; \lambda)$ and $g_\lambda = i f_{1-\lambda}$. The points (3) correspond in $\mathbf{C}$ to "elliptic logarithms" u_λ, v_λ, say; and since $\mathbf{C} = \mathbf{R} f_\lambda + \mathbf{R} g_\lambda$ there are real functions $p_\lambda, q_\lambda, r_\lambda, s_\lambda$ with

$$u_\lambda = p_\lambda f_\lambda + q_\lambda g_\lambda, \quad v_\lambda = r_\lambda f_\lambda + s_\lambda g_\lambda.$$

As λ moves, the locus of $z_\lambda = (p_\lambda, q_\lambda, r_\lambda, s_\lambda)$ has real dimension two in $\mathbf{R}^4$ and is in fact a (sub-)analytic surface Z. The torsion in $\mathbf{C}$ is $\mathbf{Q}f_\lambda + \mathbf{Q}g_\lambda$, and so our particular λ gives a point of $Z \cap \mathbf{Q}^4$. It is even in $Z \cap \frac{1}{N}\mathbf{Z}^4$ if the torsion order divides N. Such points cannot be very numerous: in Wilkie's talks we saw that the cardinality

$$|Z^{trans} \cap \frac{1}{N}\mathbf{Z}^4 \cap K| \leq c(\epsilon, Z, K)N^\epsilon$$

for a certain subset Z^{trans} of Z and any compact K (maybe this could be eliminated using o-minimality) and any $\epsilon > 0$. See also Pila [Pil] and Pila-Wilkie [PW]. It may be very difficult to write down $c(\epsilon, Z, K)$ in an effective way.

Here it is possible to show that $Z^{trans} = Z$; this is a concealed algebraic independence result as in Pila's talks, for which the hard (Hodge-theoretical) work was done by André [An].

We deal with K using bounded height: more later. We get at most $c(\epsilon)N^\epsilon$ points z_λ. An easy argument with a faint flavour of zero-estimates of the type used in transcendence theory leads to at most $c(\epsilon)N^\epsilon$ values λ. But right now we don't know any upper bound for N.

In fact it is easy to see that these values λ all lie in $\overline{\mathbf{Q}}$; further any given λ yields $D = [\mathbf{Q}(\lambda) : \mathbf{Q}]$ in all by conjugation λ^σ (compare also Habegger's talks). We deduce

$$D \leq c(\epsilon)N^\epsilon. \tag{7.7}$$

But there are also lower bounds for D. If we go back to the original problem of $\lambda, 1 - \lambda$, now with say λ of exact order N_1 then of course $D = \phi(N_1)$ for the Euler function, and this is classically known to be at least $c_1(\epsilon)N_1^{1-\epsilon}$ (now all constants are assumed positive). For our problem the analogue is that if $(2, \sqrt{4 - 2\lambda})$ in (3) has exact order N_1 then a famous theorem of Serre [Se1] implies even

$$D \geq C_1 N_1^2 \tag{7.8}$$

(which is classical in the case of complex multiplication); but here the elliptic curve depends on λ and therefore so does C_1. Furthermore it is not so easy to calculate this dependence. The work [MW2] (based on transcendence among other things) applies only if N_1 is prime, an assumption we cannot afford. It was extended to arbitrary N_1 by Zywina [Zy], but only for an elliptic curve defined over $\mathbf{Q}$, which we also cannot assume here. Very recently Lombardo [L] has extended the field of definition to $\overline{\mathbf{Q}}$; in a first version the dependence on D was not quite good enough for application here, but he has since fixed this. There is also a dependence on the absolute height $h(\lambda)$ of λ. Fortunately a theorem of Silverman [Si] implies that this height is bounded above by an absolute constant. Combining everything leads to $D \geq cN_1^\delta$ now with c absolute. Here δ is less than 10^{-10}.

For effectivity purposes it will probably be very convenient to have a bigger δ. This arises from a more direct application of the transcendence methods, starting with an exponent smaller than 2 in (8). Then the very precise version [Davi2] due to David yields $D \geq cN_1^{1/2}$.

Similarly $D \geq cN_2^{1/2}$ for the exact order N_2 of $(3, \sqrt{18 - 6\lambda})$ in (3). Taking N now as the exact order, we have $N = \operatorname{lcm}(N_1, N_2) \leq N_1 N_2$, and it follows that $D \geq cN^{1/4}$. Comparing this with (7), we see that it suffices to choose $\epsilon = \frac{1}{5}$ to get an absolute bound for $D = [\mathbf{Q}(\lambda) : \mathbf{Q}]$. Combined with the absolute bound on $h(\lambda)$, this gives by a well-known result of Northcott (see below) the required finiteness.

As height bounds were not much mentioned in the other talks, we sketch here a proof that

$$h(\lambda) \leq 6$$

for the absolute height

$$h(\lambda) = \frac{1}{D} \log \left(|a_0| \prod_\sigma \max\{1, |\lambda^\sigma|\} \right), \qquad (7.9)$$

where $a_0 \lambda^D + \cdots = 0$ is the minimal equation for λ over $\mathbf{Z}$. All we use is $N_1 P_1 = 0$ for $P_1 = (2, \sqrt{4 - 2\lambda})$, where the value of $N_1 \geq 1$ is now irrelevant.

For any algebraic λ and $P = (x, y)$ on E_λ with algebraic x, y we can reasonably define $h(P) = h(x)$ (but in Habegger's talks it was $h(x)/2$), as y is determined by x and λ. For example $h(P_1) = \log 2$. The Néron-Tate height $\hat{h}(P)$ is defined for example by

$$\hat{h}(P) = \lim_{k \to \infty} \frac{h(2^k P)}{4^k},$$

and $|\hat{h}(P) - h(P)|$ is bounded above independently of P. Explicit bounds for Weierstrass elliptic curves are practically classical, but I calculated for Legendre

$$|\hat{h}(P) - h(P)| \leq \frac{5}{3} h(\lambda) + c \qquad (7.10)$$

with explicit c absolute.

Further $\hat{h}(P) = 0$ if and only if P is torsion.

We look at $4P_1$ on E_λ. With the help of classical duplication formulae we find

$$4P_1 = \left(\frac{A(\lambda)}{B(\lambda)}, *\right)$$

with A, B in $\mathbf{Z}[t]$ of degrees 8 and 7 respectively. In fact

$$A = t^8 - 160t^7 + 7104t^6 - 57344t^5 + 206336t^4 - 401408t^3$$
$$+ 442368t^2 - 262144t + 65536,$$
$$B_4 = -288t^7 + 3648t^6 - 17408t^5 + 38912t^4 - 40960t^3 + 16384t^2.$$

Now $h(\lambda^8) = 8h(\lambda)$ (not transparent from (9) by the way) and similarly one can show, after a bit of effort with resultants, that

$$h(4P_1) \geq 8h(\lambda) - c'$$

with c' absolute.

On the other hand (10) gives

$$h(4P_1) \leq \hat{h}(4P_1) + \frac{5}{3}h(\lambda) + c = \frac{5}{3}h(\lambda) + c$$

so

$$h(\lambda) \leq \frac{c + c'}{19/3} < 6$$

with some extra computation.

Here the Northcott result just mentioned becomes clear: if $h(\lambda)$ and D are both bounded above, then so are $|a_0|$ and the $|\lambda^\sigma|$ in (9), and then so are the absolute values of the coefficients in the minimal polynomial $a_0 \prod_\sigma (t - \lambda^\sigma)$.

This completes the sketch for (3) and E_λ^2. The general curve in E_λ^2 was treated in [MZ2], and more general products like $E_\lambda \times E_{-\lambda}$ in [MZ3]. For A_λ as in [MZ4] and [MZ5] there are several extra technicalities. The results of André and Silverman apply also to the abelian situation. But despite the enormous advances by Serre in [Se2], still the extensions of [Se1] seem to be less clear-cut, even for powers of a fixed prime, let alone effective. But once more a transcendence approach succeeds, and we use David's result in [Davi1] pre-dating [Davi2]. In fact this result seems to require that the value λ is such that A_λ is simple. At first sight this looks like a problem. At second sight one suspects that such λ are probably rare, possibly controlled by conjectures of André-Oort-Pink-Zilber type (see [Pin] and [Zi] for example). At third sight one realizes that such conjectures are not yet proved. But finally by going back into the proof in [Davi1] to winkle out the "obstruction subgroup" in the zero-estimate, one sees that some easy tricks from the geometry of numbers (as in

[MW1] for example) suffice. One ends up with $D \geq cN^\delta$ with a ridiculously small exponent δ (depending on the dimension of A_λ). One could also use [MW3] to factorize the non-simple A_λ, but then the exponent would be even smaller. Probably recent work of Gaudron and Rémond [GR] would give more reasonable values.

Now for the applications of these results, denoted by (I) and (II) below.

(I) All know that Pell's equation

$$x^2 - dy^2 = 1, \quad y \neq 0$$

is solvable over $\mathbf{Z}$ provided $d > 0$ in $\mathbf{Z}$ is not a square. Moving to the polynomial ring $\mathbf{C}[t]$, by now a knee-jerk reaction, especially in view of "abcology", we consider

$$X^2 - DY^2 = 1, \quad Y \neq 0 \tag{7.11}$$

with D in $\mathbf{C}[t]$ not a square, surely easier. But in fact it is much more difficult to describe the set of D for which there is solvability over $\mathbf{C}[t]$. One can easily see that the degree m of D must be even. Now we proceed systematically.

$\underline{m = 2}$: there is always solvability. Thus for $D = at^2 + bt + c$ we can take

$$X = \frac{2at + b}{\sqrt{b^2 - 4ac}}, \quad Y = \frac{2\sqrt{a}}{\sqrt{b^2 - 4ac}}.$$

$\underline{m = 4}$: there is not always solvability. For example not for $D = t^4 + t + 1$. And in the family $D = t^4 + t + \lambda$ we have solvability exactly when λ lies in a certain countable subset of $\mathbf{C}$. In fact precisely when the point $(0, 1)$ is torsion on $y^2 = x^3 - 4\lambda x + 1$. This is essentially classical (Abel [Ab], Chebychev [C1],[C2]). In fact the set is infinite (which is not classical – for several proofs see [Za] pages 92,93 for example), as one might guess from its element (with its six conjugates)

$$\lambda = \frac{\sqrt[3]{2\sqrt{2} - 2}}{2}$$

where

$$X = \frac{(4 - 32\lambda^3)t^5 - (4\lambda - 16\lambda^4)t^4 + 4\lambda^2 t^3 + (3 - 28\lambda^3)t^2 - 8\lambda^4 t + 8\lambda^5}{32\lambda^8},$$

$$Y = \frac{(4 - 32\lambda^3)t^3 - (4\lambda - 16\lambda^4)t^2 + 4\lambda^2 t + (1 - 12\lambda^3)}{32\lambda^8}.$$

$\underline{m = 6}$: there is rarely solvability. For example there are only finitely many λ in $\mathbf{C}$ for which solvability holds for $D = t^6 + t + \lambda$. This is proved in [MZ4] using the general result on A_λ described above. In fact here A_λ is the Jacobian of the hyperelliptic curve $s^2 = t^6 + t + \lambda$ of genus 2, or better a complete non-singular model

$$s^2 = t_3^2 + t_0 t_1 + \lambda t_0^2, \quad t_0 t_2 = t_1^2, \quad t_0 t_3 = t_1 t_2.$$

The curve inside A_λ is the locus of the divisor $\Delta_\lambda = \infty_\lambda^+ - \infty_\lambda^-$ as λ varies, where $\infty_\lambda^\pm$ are the two places at infinity. When (11) holds we write it as $f_\lambda^+ f_\lambda^- = 1$ with the functions $f_\lambda^\pm = X \pm sY$ to see that their divisors are multiples of Δ_λ thus giving a torsion point.

There are actually some λ; for example $\lambda = 0$ with

$$X = 2t^5 + 1, \quad Y = 2t^2.$$

$m \geq 8$: even rarer. For example with the family $D = d_0(\lambda) t^m + \cdots + d_m(\lambda)$ in $\overline{\mathbf{Q}}[\lambda][t]$, say for safety identically squarefree, there is solvability for infinitely many λ in $\overline{\mathbf{Q}}$ only if the analogous Jacobian, now an abelian variety of dimension $\frac{m-2}{2} \geq 3$, contains an elliptic curve. This also follows from the A_λ result in [MZ5].

Incidentally, if we want to go beyond squarefree, then we can use the result of [BMPZ] on multiplicative extensions (4). Thus for $D = t^2(t^4 + t + \lambda)$ we get at most a finite set, despite the infinite set for $t^4 + t + \lambda$. And also for $D = t^3(t^3 + t + \lambda)$ using the additive extensions (6).

(II) This concerns the old problem of "integrating in elementary terms" (see for example the article [R] by Risch). By the way, the integration may be elementary but it need not be easy (just as for some proofs), as a wonderful example

$$\int \frac{\sqrt{1+t^4}}{1-t^4} \, dt = -\frac{1}{4}\sqrt{2} \log \left(\frac{\sqrt{2}t - \sqrt{1+t^4}}{1-t^2} \right)$$
$$- \frac{i}{4}\sqrt{2} \log \left(\frac{i\sqrt{2}t + \sqrt{1+t^4}}{1+t^2} \right) \tag{7.12}$$

due to Euler shows. Not only can my Maple (version 9) not do the integration on the left-hand side, but it cannot even check the result by differentiating the right-hand side. Actually Euler's version was

$$\frac{1}{4}\sqrt{2} \log \left(\frac{\sqrt{2}t + \sqrt{1+t^4}}{1-t^2} \right) + \frac{1}{4}\sqrt{2} \arcsin \left(\frac{\sqrt{2}t}{1+t^2} \right)$$

thus staying over $\mathbf{R}$.

We give some more examples in the above hyperelliptic context.

$m = 2$: now $\int \frac{dt}{\sqrt{D}}$ is always integrable – see any engineer's handbook of indefinite integrals.

$\underline{m = 4}$: now $\int \frac{dt}{\sqrt{t^4+t+\lambda}}$ is integrable if and only if $256\lambda^3 = 27$; then we can reduce it to

$$\int \frac{dt}{(t + 4\lambda/3)\sqrt{t^2 + \mu t + \nu}}$$

and run to the handbook.

$\underline{m = 6}$: now the same methods show that $\int \frac{dt}{\sqrt{t^6+t+\lambda}}$ is integrable only if $46656\lambda^5 = 3125$, and with a bit more effort never.

However up to now all that is in fact relatively easy, and not at the same level as Pell. But

$$\int \frac{dt}{t\sqrt{t^4 + t + \lambda}}$$

is integrable if and only if λ lies in a certain finite set. Oddly enough the proof does not use multiplicative extensions as for $t^2(t^4 + t + \lambda)$ above but rather Schmidt's result [Sc] on additive extensions (6). Incidentally, he has made such results effective, using among other things a version [Ma] of the original result of Bombieri-Pila [BP] obtained as in Wilkie's talks with the Siegel Lemma. For example he shows that there are at most $e^{e^{e^{e^5}}}$ complex values of λ for which

$$\int \frac{dt}{(t - 2)\sqrt{t(t - 1)(t - \lambda)}}$$

is integrable. This is related to (3): thus integrability implies that $(2, \sqrt{4 - 2\lambda})$ is torsion on E_λ. But the converse fails, so we cannot deduce infinitely many λ. In fact we get a torsion point even on a suitable G_λ as in (6), so we may conclude finiteness.

And also by [MZ5]

$$\int \frac{dt}{t\sqrt{t^6 + t + \lambda}}$$

is integrable at most on a finite set; but no-one knows how to make this effective. Here we use A_λ as above, but now with the locus of $\Gamma_\lambda = P_\lambda^+ - P_\lambda^-$, where $P_\lambda^\pm = (0, \pm\sqrt{\lambda})$; this time the torsion property arises from a classical criterion of Liouville, which implies that the integral, if elementary, must involve a single $\log g_\lambda$ (as opposed to (12) with a pair). Now differentiation (without Maple) gives $dt/(ts) = cdg_\lambda/g_\lambda$ for c in $\mathbf{C}$, from which we see that the divisor of g_λ is a multiple of Γ_λ.

These examples support an assertion of James Davenport [Dave] from 1981.

I thank Gareth Jones for comments on an earlier version.

References

[Ab]   N.H. Abel, *Über die Integration der Differential-Formel $\rho dx/\sqrt{R}$, wenn R und ρ ganze Funktionen sind*, J. für Math. (Crelle) **1** (1826), 185–221.

[An]   Y. André, *Mumford-Tate groups of mixed Hodge structures and the theorem of the fixed part*, Compositio Math. **82** (1992), 1–24.

[Be]   D. Bertrand, *Special points and Poincaré bi-extensions; with an Appendix by Bas Edixhoven*, ArXiv 1104.5178v1.

[BMPZ] D. Bertrand, D. Masser, A. Pillay, and U. Zannier, *Relative Manin-Mumford for semi-abelian surfaces*, submitted for publication.

[BP]   E. Bombieri and J. Pila, *The number of integral points on arcs and ovals*, Duke Math. J. **59** (1989), 337–357.

[C1]   P. Chebychev, *Sur l'intégration des différentielles qui contiennent une racine carée d'un polynome du troisième ou du quatrième degré*, J. Math. Pures Appl. **2** (1857), 168–192.

[C2]   P. Chebychev, *Sur l'intégration de la différentielle* $\dfrac{x+A}{\sqrt{x^4+\alpha x^3+\beta x^2+\gamma x+\delta}}$, J. Math. Pures Appl. **9** (1864), 225–246.

[Dave] J.H. Davenport, *On the integration of algebraic functions*, Lecture Notes in Computer Science **102**, Springer-Verlag, Berlin Heidelberg New York 1981.

[Davi1] S. David, *Fonctions theta et points de torsion des variétés abéliennes*, Compositio Math. **78** (1991), 121–160.

[Davi2] S. David, *Points de petite hauteur sur les courbes elliptiques*, J. Number Theory **64** (1997), 104–129.

[GR]   É. Gaudron and G. Rémond, *Théorème des périodes et degrés minimaux d'isogénies*, to appear in Commentarii Math. Helv. (39 pages).

[H]   M. Hindry, *Autour d'une conjecture de Serge Lang*, Inventiones Math. **94** (1988), 575–604.

[L]   D. Lombardo, *Bounds for Serre's open image theorem for elliptic curves over number fields*, preprint 2014 (36 pages).

[Ma]   D. Masser, *Rational values of the Riemann zeta function*, J. Number Theory **131** (2011), 2037–2046.

[MW1]   D. Masser and G. Wüstholz, *Periods and minimal abelian subvarieties*, Ann. of Math. **137** (1993), 407–458.

[MW2]   D. Masser and G. Wüstholz, *Galois properties of division fields of elliptic curves*, Bull. London Math. Soc. **25** (1993), 247–254.

[MW3]   D. Masser and G. Wüstholz, *Factorization estimates for abelian varieties*, Publ. Math. IHES **81** (1995), 5–24.

[MZ1]   D. Masser and U. Zannier, *Torsion anomalous points and families of elliptic curves*, Amer. J. Math. **132** (2010), 1677–1691.

[MZ2]   D. Masser and U. Zannier, *Torsion points on families of squares of elliptic curves*, Math. Annalen **352** (2012), 453–484.

[MZ3]   D. Masser and U. Zannier, *Torsion points on families of products of elliptic curves*, Advances in Math **254** (2014), 116–133.

[MZ4]   D. Masser and U. Zannier, *Torsion points on families of abelian surfaces and Pell's equation over polynomial rings* (with Appendix by V. Flynn), submitted for publication.

[MZ5]   D. Masser and U. Zannier, *Torsion points on families of abelian varieties, Pell's equation, and integration in elementary terms*, in preparation.

[Pil]   J. Pila, *Integer points on the dilation of a subanalytic surface*, Quart. J. Math. **55** (2004), 207–223.

[PW]   J. Pila and A. Wilkie, *The rational points of a definable set*, Duke Math. J. **133** (2006), 591–616.

[PZ]   J. Pila and U. Zannier, *Rational points in periodic analytic sets and the Manin-Mumford conjecture*, Rendiconti Lincei Mat. Appl. **19** (2008), 149–162.

[Pin]   R. Pink, *A common generalization of the conjectures of André-Oort, Manin-Mumford, and Mordell-Lang*, manuscript dated 17th April 2005 (13 pages).

[R]   R.H. Risch, *The problem of integration in finite terms*, Trans. Amer. Math. Soc. **139** (1969), 167–189.

[Sc]   H. Schmidt, Ph.D. thesis (Basle), in preparation.

[Se1]   J.-P. Serre, *Propriétés galoisiennes des points d'ordre fini des courbes elliptiques*, Inventiones Math. **15** (1972), 259–331.

[Se2]   J.-P. Serre, *Résumé des cours au Collège de France de 1985-1986*, Collected papers IV 1985-1998, Springer-Verlag, Berlin 2000.

[Si]   J.H. Silverman, *Heights and the specialization map for families of abelian varieties*, J. reine angew. Math. **342** (1983), 197–211.

[Za]   U. Zannier, *Some problems of unlikely intersections in arithmetic and geometry*, Annals of Math. Studies **181**, Princeton 2012.

[Zi]   B. Zilber, *Exponential sums equations and the Schanuel conjecture*, J. London Math. Soc. **65** (2002), 27–44.

[Zy]   D. Zywina, *Bounds for Serre's open image theorem*, ArXiv e-prints, 1102.4656, February 2011 (16 pages).

8

Improving the bound in the Pila-Wilkie
theorem for curves

G. O. Jones

Suppose that $f : (a,b) \to \mathbb{R}$ is an analytic function definable in an o-minimal expansion of the real field, and suppose that f is transcendental, that is that there is no nonzero polynomial P such that $P(t,f(t))$ vanishes identically. The Pila-Wilkie theorem applied to the graph of f says that for all $\varepsilon > 0$ there is a $c > 0$ such that for all $H \geq 1$ there are at most

$$cH^{\varepsilon}$$

rationals q in (a,b) of height at most H such that $f(q)$ is also a rational of height at most H. See one of Wilkie's contributions to this volume for a discussion of this result, and its proof. The analyticity isn't necessary, and certainly isn't true piecewise of definable functions in general (indeed, they needn't even be piecewise infinitely differentiable, see [12]). But all the functions we will meet later are (possibly piecewise) analytic, so we may as well assume it from the beginning.

It is reasonable to ask whether the cH^{ε} bound in this result can be improved, say to a bound of the form $c(\log H)^n$ for some $c, n > 0$. In fact, this sort of improvement is not possible. Constructions due to Surroca [19, 20] and to Bombieri and Pila [17] show the following. Suppose that $\varepsilon : [1, \infty) \to \mathbb{R}$ is strictly decreasing and tends to 0. Then there is a transcendental real analytic function f on $[0, 1]$ (that is, f has an analytic extension to a neighbourhood of $[0, 1]$) and an increasing sequence $H_1, H_2, \ldots$ of positive integers such that for each positive integer i there are at least

$$H_i^{\varepsilon(H_i)}$$

O-Minimality and Diophantine Geometry, ed. G. O. Jones and A. J. Wilkie. Published by Cambridge University Press. © Cambridge University Press 2015.

rational points $(q, f(q))$ on the graph of f of height at most H_i. Taking $s(t) = \frac{1}{\sqrt{\log(t+1)}}$ the function f provided by this construction fails to satisfy the improvement of Pila-Wilkie to a bound of the form $c(\log H)^n$. Surroca's examples even have continuations to entire functions and are such that for any algebraic α and any natural number i the number $f^{(i)}(\alpha)$ lies in the field $\mathbb{Q}(\alpha)$. See also [4] for further constructions of examples satisfying reasonably nice analytic properties.

The constructions leading to these examples have a somewhat artificial feel, so we modify our question to ask whether the bound can be improved when f is in some way natural. Of course, this new question has the additional feature that it is so vague it can never be fully answered. Nonetheless, given the examples, it is the best we can hope for. There are at least two reasonable directions in which we might pursue positive answers to the new question. First, we could assume that our function f is definable in some particular o-minimal structure, for example $\mathbb{R}_{\exp}$, the expansion of the real field by the exponential function. Here, Wilkie has conjectured that the bound in the Pila-Wilkie theorem can be improved to one of the form $c(\log H)^n$ (his conjecture is that this improvement is possible for *any* definable set, but we stick with curves). The second possibility is perhaps more a subcase of the first, and is to just assume that f is some particular concrete function. For example we could take f to be exp, in which case we're fine (by Hermite and Lindemann) or f could be x^α (still fine, by the six exponentials theorem, see [11]), or perhaps the Riemann zeta function restricted to a bounded interval, or the gamma function, similarly restricted. For these latter two examples, we'll see later the best results that are known.

At present, all proofs of improvements along the lines we have outlined start with something like the following.

Proposition   *Suppose that $f : [0,1] \to [0,1]$ is analytic. There is a constant $c > 0$ depending on f with the following property. For all $H > e$ there is a nonzero polynomial P of (total) degree at most $c \log H$ such that if $(q, f(q))$ is a rational point of height at most H then $P(q, f(q)) = 0$.*

Here is a brief sketch of one proof of this result. Throughout this survey, we use c to denote positive constants which may differ at each occurrence. Let d be a positive integer (later we set d to be $[c \log H]$) and $D = \frac{(d+1)(d+2)}{2}$ the number of monomials in X, Y of degree at most d. Suppose that (a, b) is an interval in $(0, 1)$ such that there are rational points $(q_1, f(q_1)), \ldots, (q_D, f(q_D))$ of height at most H with $q_1 < \cdots < q_D$ and each q_i in (a, b). If there are not this many points of height H then we can certainly find a polynomial of degree at most d vanishing at them. Suppose that our points q_i are such that for any nonzero polynomial P of degree at most d there is some i with $P(q_i, f(q_i)) \neq 0$. Our aim is to show that then $b - a$ is not too small.

We first show that we must have

$$\Delta = \det\left(m_j(q_i, f(q_i))\right)$$

nonzero, where $m_1, \ldots, m_D$ is an enumeration of the monomials in X, Y of degree at most d. So suppose that $\Delta = 0$, so that our matrix has rank, r say, at most $D - 1$. Fix an $r \times r$ minor of rank r. For convenience we suppose that this minor is

$$\left(m_j(q_i, f(q_i))\right)_{1 \leq i,j \leq r}.$$

Consider the polynomial $Q(X, Y)$ given by

$$\det \begin{pmatrix} m_1(q_1, f(q_1)) & \cdots & m_1(q_r, f(q_r)) & m_1(X, Y) \\ \vdots & \vdots & \vdots & \vdots \\ m_{r+1}(q_1, f(q_1)) & \cdots & m_{r+1}(q_r, f(q_r)) & m_{r+1}(X, Y) \end{pmatrix}. \qquad (8.1)$$

Then for $i > r$ we have $Q(q_i, f(q_i)) = 0$ since otherwise our original matrix would have rank greater than r. And for $i \leq r$ we have $Q(q_i, f(q_i)) = 0$ since then the matrix in ((8.1)) has a repeated column. So Q vanishes at all our points, which is a contradiction as Q has degree at most d. Hence we do have $\Delta \neq 0$ as claimed.

Now, all the entries in our matrix are rationals of height at most H and so Δ is certainly rational and a calculation shows that $H^{2dD}\Delta$ is an integer. As this integer is nonzero, the 'Fundamental Theorem of Transcendence' says that

$$H^{2dD}|\Delta| \geq 1. \qquad (8.2)$$

We then use Proposition 2 from [3, page 341] which says that

$$\Delta = \det V \det L \qquad (8.3)$$

where V is the Vandermonde matrix

$$\begin{pmatrix} 1 & q_1 & \cdots & q_1^{D-1} \\ \vdots & \vdots & \vdots & \vdots \\ 1 & q_D & \cdots & q_D^{D-1} \end{pmatrix}$$

and L is the matrix

$$\begin{pmatrix} \phi_1(\xi_{1,1}) & \cdots & \frac{\phi_1^{(D-1)}(\xi_{D,1})}{(D-1)!} \\ \vdots & \vdots & \vdots \\ \phi_D(\xi_{1,D}) & \cdots & \frac{\phi_D^{(D-1)}(\xi_{D,D})}{(D-1)!} \end{pmatrix}$$

with $\phi_j(t) = m_j(t, f(t))$ and suitable points $\xi_{i,j}$ in the interval (a, b). The point of (8.3) is that we can use it to find an upper bound on $|\Delta|$. To handle $|\det L|$ we use the fact that f is analytic on a neighbourhood of $[0, 1]$ and so there is an $A > 0$ such that

$$\left| f^{(j)}(t) \right| \leq j! A^j$$

for all j. The Vandermonde matrix V has determinant

$$\det V = \prod_{1 \leq i < i' \leq D} (q_{i'} - q_i)$$

which is at most $|b - a|^{D(D-1)/2}$. Combining these we get an upper bound on $|\Delta|$ and then (8.2) gives a lower bound on the length of the interval (a, b). So we can then cover $(0, 1)$ with shorter intervals, and, by the above, for each of these shorter intervals there is a curve of degree at most d which contains all the points lying above the interval. Carrying out all the computations that we've skipped shows that we need at most

$$6AH^{\frac{8}{d+3}}$$

intervals. So we can multiply the resulting polynomials together and use d around $[c \log H]$ to find a final polynomial of degree at most $c \log H$ (with a different c).

This sort of argument goes back to the 1989 paper by Bombieri and Pila [3] and the particular version just sketched is from [15] by Pila. (The proposition is 2.4 of [15] with $\phi_1(t) = t, \phi_2(t) = f(t)$, B there equal to our A, $C = 0$ and $d = [5 \log H]$.) The same proposition can also be proved using classical transcendence techniques. We first use Siegel's lemma (see for instance [1]) to find a polynomial P vanishing at some of the given points. Then, if it doesn't vanish at another of the points then the height bound and the construction of the polynomial tell us that the polynomial can't be too small at this point. We can now appeal to complex analysis (our function has a continuation to a complex analytic function on a complex neighbourhood of $[0, 1]$). More specifically we can use the maximum modulus theorem to show that as the function $P(z, f(z))$ has quite a few zeros it must be fairly small at the given point. In fact, it must be small enough that the lower bound above means it must indeed vanish at our given point. For details of this argument to prove a stronger version of the proposition see Proposition 2 in [14] and its proof.

Once we have the proposition we can turn to the second part of the argument. We have our transcendental analytic function $f : [0,1] \to [0,1]$ (if the codomain is a larger subset of the reals then we can divide by a sufficiently large integer, and this will only affect constants) and the proposition means that we'll be done if we can find suitably good bounds on the number of zeros of functions of the form $P(t,f(t))$ for nonzero polynomials P in terms of the degree of P. In the case that f is definable in $\mathbb{R}_{\exp}$ we have the following wonderful theorem, due to Khovanskii [10], to help us at this point.

Theorem *Suppose that $Q_1,\ldots,Q_k$ are polynomials of degree at most d in $2n$ variables with real coefficients. Then there is a constant c depending only on n such that the set*

$$\{x \in \mathbb{R}^n : Q_1(x,e^{x_1},\ldots,e^{x_n}) = \cdots = Q_k(x,e^{x_1},\ldots,e^{x_n}) = 0\}$$

has at most

$$cd^{2n}$$

connected components.

This theorem, proved in the 1970s, has been very influential in the development of o-minimality. For the proof see either Khovanskii's book [10] or Marker's notes [13] (which do not contain the explicit bound).

How does this help? Suppose that our function is not only definable in $\mathbb{R}_{\exp}$ but picks out part of the zero set of a system as in the theorem with $n = 2$. More precisely, suppose that the graph of f is contained in

$$\{(x,y) \in \mathbb{R}^2 : Q(x,y,e^x,e^y) = 0\}$$

with Q a nonzero polynomial. Let P be the polynomial provided by the proposition, so P has degree at most $c \log H$ and if $(q,f(q))$ is a rational point of height at most H then $P(q,f(q)) = 0$. Then the rational points of height at most H are contained in the set

$$\{(x,y) \in \mathbb{R}^2 : Q(x,y,e^x,e^y) = P(x,y) = 0\}.$$

Since f is transcendental, these points are actually connected components of this set and so we can appeal to Khovanskii's theorem to conclude that there are at most

$$c(\log H)^4$$

points. Although this looks like a very special case it is not so far from the general case. Wilkie proved in 1991 [23] that the theory of $\mathbb{R}_{exp}$ is model complete. This means that if $X \subseteq \mathbb{R}^n$ is any definable set in $\mathbb{R}_{exp}$ then there is some nonnegative integer m and a real polynomial Q in $2n + 2m$ variables such that

$$X = \pi \left(\{ (x, y) \in \mathbb{R}^{n+m} : Q\left(x, y, e^{x_1}, \ldots, e^{x_n}, e^{y_1}, \ldots, e^{y_m} \right) = 0 \} \right), \quad (8.4)$$

where π is the projection onto the first n coordinates. Note that o-minimality of $\mathbb{R}_{exp}$ follows immediately from Wilkie's theorem by Khovanskii's theorem. Using Wilkie's theorem, we can argue more or less as above to show that if $f : [0, 1] \to \mathbb{R}$ is a transcendental analytic function definable in $\mathbb{R}_{exp}$ then the bound in the Pila-Wilkie theorem applied to the graph of f can be improved to $c(\log H)^n$ for some c and n depending on f. To go further, beyond the case that f is analytic on a compact interval, we need some more definability theory. Using some of the ingredients of Wilkie's proof, the representation of definable sets in (8.4) can be refined. Roughly, we can assume that X can be written piecewise as projections as in (8.4) in which the variety being projected is smooth and the projection has maximal rank. Here is a precise statement for definable functions $f : (a, b) \to \mathbb{R}$. There are $a = a_0 < a_1 < \cdots < a_{k+1} = b$ such that on each interval (a_i, a_{i+1}) the following holds. There exist $m \geq 0, \phi_1, \ldots, \phi_m : (a_i, a_{i+1}) \to \mathbb{R}$ and polynomials $Q_1, \ldots, Q_{m+1}$ in $2m + 2$ variables (all depending on i, but we suppress this) such that after putting $F_j(x, y, z_1, \ldots, z_m) = Q_j(x, y, z, e^x, e^y, e^{z_1}, \ldots, e^{z_m})$ we have

$$F_1(t, f(t), \phi(t)) = 0$$

$$\vdots \qquad\qquad (8.5)$$

$$F_{m+1}(t, f(t), \phi(t)) = 0$$

and

$$\det \begin{pmatrix} \frac{\partial F_1}{\partial y} & \frac{\partial F_1}{\partial z_1} & \cdots & \frac{\partial F_1}{\partial z_m} \\ \vdots & \vdots & \vdots & \vdots \\ \frac{\partial F_{m+1}}{\partial y} & \frac{\partial F_{m+1}}{\partial z_1} & \cdots & \frac{\partial F_{m+1}}{\partial z_m} \end{pmatrix} (t, f(t), \phi(t)) \neq 0$$

for t in (a_i, a_{i+1}), where $\phi(t) = (\phi_1(t), \ldots, \phi_m(t))$. We say that f is *implicitly definable* on (a_i, a_{i+1}), and so the result is that any definable function is piecewise implicitly definable. This is proved by combining Wilkie's model completeness results with part of its proof. See also [9] for another proof, assuming model completeness, and a more general setting. Note that since definable functions in $\mathbb{R}_{exp}$ are piecewise implicitly defined they are also piecewise analytic.

To go beyond the compact case we also need improvements to the diophantine part of the proof. Note that in the proof of the proposition the analyticity of f is used to get bounds on derivatives. If instead we assumed that f is smooth and satisfies weaker bounds we would still get somewhere. And we don't need infinite differentiability, only differentiability to a certain order. We can get beyond the proposition by leaving in parameters for the order of differentiability, the bounds on derivatives and the degree of the polynomial we're aiming for (which in the end will again be $\log H$). We introduce some notation, following Pila's presentation in [16]. Let I be an interval (of any type) and suppose that $f : I \to \mathbb{R}$ has k continuous derivatives. Given a positive L, set $T_{L,0} = 1$ and for $k > 0$ put

$$T_{L,k} = \max_{1 \le i \le k} \left(1, \sup_{x \in I} \left(\frac{|f^{(i)}(x)| L^{i-1}}{i!} \right)^{1/i} \right),$$

so $T_{L,k}$ may be infinite. Pila proved the following (see [18, 2.5] and [16, 2.2]).

Proposition *Let $d \ge 1, D = (d+1)(d+2)/2$ and suppose that $H \ge 1$ and $L \ge 1/H^3$. Suppose that I is an interval of length at most L and that $f : I \to \mathbb{R}$ has $D - 1$ continuous derivatives, with $|f'(x)| \le 1$ for x in I. Then the set of rational points of height at most H lying on the graph of f is contained in the union of at most*

$$6 T_{L,D-1}(f) L^{\frac{8}{3(d+3)}} H^{\frac{8}{d+3}}$$

real algebraic curves of degree at most d.

From this proposition, by a very careful recursion, Pila proved the following (see [16, 3.2]).

Proposition *Let d, D be as in the proposition above and suppose that $H \ge e$ and $L > 1/H^2$. Suppose that I is an interval of length at most L and that $f : I \to \mathbb{R}$ has D continuous derivatives, with $|f'(x)| \le 1$ on I and such that each $f^{(i)}$ is either nonvanishing on the interior of I or is identically zero. Then the set of rational points of height at most H lying on the graph of f is contained in at most*

$$66 D \left(L H^3 \right)^{\frac{8}{3(d+3)}} \log \left(e L H^2 \right)$$

real algebraic curves of degree at most d.

We can now prove Wilkie's conjecture in the case that our set is the graph of an analytic transcendental function $f : I \to \mathbb{R}$ on an interval I. We no longer suppose that I is compact. It is easy to deduce the conjecture for curves from this special case. First, by the discussion of implicit definability above we can assume that f is implicitly defined, as in (8.5). Let $k \geq 0$. We first divide I into intervals on which $f'(x)$ is either ≤ -1, in $(-1, 1)$ or ≥ 1. We then further divide into intervals on which either f (in the second case) or its inverse (in the other two cases) has nonvanishing derivatives up to order k. Using Khovanskii's theorem and differentiating the equations in (8.5) we find that this can be achieved with at most

$$ck^n$$

intervals, where c and n depend only on the representation of f in (8.5). Now let $d \geq 1$ and $D = (d+1)(d+2)/2$ as before. Since a rational of height at most H lies in the interval $[-H, H]$ we can assume that our original interval has length at most $2H$. We subdivide as above, up to $k = D$. Then on each of the resulting intervals the previous proposition applies. So over each of these intervals the rational points of height at most H lie on at most

$$cDH^{\frac{32}{3(d+3)}} \log H$$

curves of degree at most d. Using Khovanskii's theorem and implicit definability again, the number of intersection points of the graph of f with a curve of degree at most d is at most

$$cd^m,$$

again with c and m depending only on the representation of f. So in total there are

$$cd^{2n+2m+2} H^{\frac{32}{3(d+3)}} \log H$$

points. Taking d to be $[\log H]$ we have the following.

Theorem   *Suppose that f is analytic on an interval I and definable in $\mathbb{R}_{\exp}$. Then there are c and k such that for all $H > e$ the number of rational points of height at most H on the graph of f is at most*

$$c(\log H)^k.$$

This theorem is due to Butler [6], and to Thomas and me [8]. The proof is based on one by Pila [16], and in fact the main work in [6] and [8] is in working out the structure of the constant c, as this plays an important role in what we know about Wilkie's conjecture for surfaces.

This completes the discussion of Wilkie's conjecture. So what about the second possibility mentioned in the introduction? Recall that this was to just consider the graph of some classical function (so no definability). The first result, not covered by transcendence results that I know of, is due to Masser [14]. This concerns the Riemann zeta function $\zeta(x) = \sum_{n \geq 1} \frac{1}{n^x}$.

Theorem   *There is a constant $c > 0$ such that for $H \geq e^e$ the number of rational points of height at most H on the graph of $\zeta|_{(2,3)}$ is at most*

$$c \left(\frac{\log H}{\log \log H} \right)^2.$$

Note that this is not covered by Wilkie's conjecture for curves: no restriction of the Riemann zeta function is definable in $\mathbb{R}_{\exp}$.

In trying to prove this result we face the same problem as before, namely bounding the number of zeros of $P(z, \zeta(z))$ for $\deg P$ at most d and $z \in (2, 3)$. To this end, Masser proves a very precise zero estimate for the entire function $(z - 1)\zeta(z)$, in terms of d and the radius of the disk on which we wish to count zeros. For details, see Masser's paper. We give below a simple proof of a weaker bound. For this proof, we make use of additional features of the polynomials that we have. Namely, our polynomial has integer coefficients, and these coefficients are not too big. Here, it is better to use the construction via complex analysis briefly mentioned above. When we inspect Masser's proof of Proposition 2 in [14] we find that the coefficients of P are integers of modulus at most

$$2(d+1)^2 H^d.$$

(See [5] for a discussion of this, although note that the T there, and in Masser's paper, is our d and the d there is something else.) The result we need from complex analysis is as follows.

Proposition   *Suppose that g is analytic on the closed disk around the origin of radius $2r$ with $|g(z)| \leq M$ for z in this disk, and suppose that $g(0) \neq 0$. Then the number of zeros of g in the closed disk around the origin of radius r is at most*

$$\frac{1}{\log 2} \left(\log M - \log |g(0)| \right). \tag{8.6}$$

This is an immediate corollary of Jensen's formula or can be proved directly using the maximum modulus theorem (see [21, Page 171]).

Let $f_1(z) = (z-1)\zeta(z)$, and let P_1 be a polynomial of degree at most $c \log H$ such that $P_1(q, f_1(q)) = 0$ whenever $(q, f_1(q))$ is a point of height at most H. Suppose that P_1 has integer coefficients, bounded as above. Using this bound, we can easily bound the M in (8.6). How do we handle the $|g(0)|$ term? What do we do if it is zero? One trick to avoid this problem is to work around $z = 2$ instead, or equivalently, to use the function $f_2(z) = (z+1)\zeta(z+2)$. Let P_2 be the associated polynomial, again with integer coefficients satisfying the bound. Now we have $f_2(0) = \pi^2/6$, and so the only way that the function $g_2(z) = P_2(z, f_2(z))$ can vanish at the origin is if the polynomial $P_2(0, Y)$ is identically 0. In this case we can divide through by an appropriate power of Y without changing any of the relevant properties of P_2. So we can assume that $P_2(0, Y)$ does not vanish identically. Then we need lower bounds on

$$|g_2(0)| = |P_2(0, \frac{\pi^2}{6})|.$$

Transcendence theorists have long been interested in lower bounds for numbers of this kind, in terms of the coefficients of $P_2(0, Y)$ and its degree. The resulting estimates are called transcendence measures. A result of Waldschmidt [22] says the following. Suppose that $Q(Y)$ is a nonzero polynomial of degree T, and let $|Q|$ denote the maximum of the absolute values of the coefficients, which we suppose is at least 16. Then

$$\log |Q(\pi)| \geq -2^{40} T \left(\log |Q| + T \log T\right) (1 + \log T).$$

Putting this into (8.6), we find a bound of the form

$$c(\log H)^3 \log \log H$$

for the number of rational points of f_2 restricted to $(0, 1)$ which immediately gives Masser's result with this bound in place of his.

All that this argument, which is due to Boxall and me [5], really used of ζ is the particular value $\zeta(2) = \pi^2/6$. So the argument also applies to other functions, for example $\zeta(x)/\pi^x$, using the fact that $\zeta(-1) = -1/12$, or it could be used for Γ, using $\Gamma(1/2) = \sqrt{\pi}$. This answers some questions left open in Masser's paper. But for Γ, Besson [2] independently proved a zero estimate similar to Masser's result for ζ. Using this he established a bound of the form $c(\log H)^2 (\log \log H)^{-1}$ for the restriction of Γ to a compact interval, say $[2, 3]$. In [5] Boxall and I also showed that slightly weaker bounds hold for the restrictions of ζ, Γ and $\zeta(x)/\pi^x$ to $(2, \infty)$. Note that the latter two examples do have infinitely many rational points. Zero estimates for Γ and

ζ were independently found earlier by Coman and Poletsky [7], motivated by problems in analysis.

The most general result I am aware of for restrictions of entire functions to compact sets requires two definitions. Suppose f is a nonconstant entire function. The *order* of f is

$$\rho = \limsup_{r \to \infty} \frac{\log \log M(r)}{\log r}$$

where $M(r)$ is the maximum of $|f(z)|$ with $|z| \leq r$. Similarly, the *lower order* is

$$\lambda = \liminf_{r \to \infty} \frac{\log \log M(r)}{\log r}.$$

Boxall and I recently proved the following.

Theorem *Suppose that f is an entire function, with finite order ρ and positive lower order λ. Then there are positive c and k such that for $H \geq e$ the number of rational points on the graph of f restricted to $[0, 1]$ is at most*

$$c(\log H)^k.$$

For the proof, see [4].

I'm grateful to Margaret Thomas and Alex Wilkie both for comments on a draft of this article and for numerous discussions on the topics covered here.

References

[1] Alan Baker. *Transcendental number theory*. Cambridge Mathematical Library. Cambridge University Press, Cambridge, second edition, 1990.

[2] Etienne Besson. Points rationnels de la fonction gamma d'euler. *Archiv der Mathematik*, 103(1):61–73, 2014.

[3] E. Bombieri and J. Pila. The number of integral points on arcs and ovals. *Duke Math. J.*, 59(2):337–357, 1989.

[4] Gareth Boxall and Gareth Jones. Rational values of entire functions of finite order. *Int. Math. Res. Not.* 2015 DOI: 10.1093/imrn/rnvo52

[5] Gareth J. Boxall and Gareth O. Jones. Algebraic values of certain analytic functions. *Int. Math. Res. Not.*, (2015)(4):1141–1158.

[6] Lee A. Butler. Some cases of Wilkie's conjecture. *Bull. Lond. Math. Soc.*, 44(4):642–660, 2012.

[7] Dan Coman and Evgeny A. Poletsky. Transcendence measures and algebraic growth of entire functions. *Invent. Math.*, 170(1):103–145, 2007.

[8] G. O. Jones and M. E. M. Thomas. The density of algebraic points on certain Pfaffian surfaces. *Q. J. Math.*, 63(3):637–651, 2012.

[9] G. O. Jones and A. J. Wilkie. Locally polynomially bounded structures. *Bull. Lond. Math. Soc.*, 40(2):239–248, 2008.

[10] A. G. Khovanskii. *Fewnomials*, volume 88 of *Translations of Mathematical Monographs*. American Mathematical Society, Providence, RI, 1991. Translated from the Russian by Smilka Zdravkovska.

[11] Serge Lang. *Introduction to transcendental numbers*. Addison-Wesley Publishing Co., Reading, Mass.-London-Don Mills, Ont., 1966.

[12] Olivier Le Gal and Jean-Philippe Rolin. An o-minimal structure which does not admit C^∞ cellular decomposition. *Ann. Inst. Fourier (Grenoble)*, 59(2):543–562, 2009.

[13] David Marker. Khovanskii's theorem. In *Algebraic model theory (Toronto, ON, 1996)*, volume 496 of *NATO Adv. Sci. Inst. Ser. C Math. Phys. Sci.*, pages 181–193. Kluwer Acad. Publ., Dordrecht, 1997.

[14] D. Masser. Rational values of the Riemann zeta function. *J. Number Theory*, 131(11):2037–2046, 2011.

[15] J. Pila. Mild parameterization and the rational points of a Pfaff curve. *Comment. Math. Univ. St. Pauli*, 55(1):1–8, 2006.

[16] J. Pila. The density of rational points on a Pfaff curve. *Ann. Fac. Sci. Toulouse Math. (6)*, 16(3):635–645, 2007.

[17] Jonathan Pila. Integer points on the dilation of a subanalytic surface. *Q. J. Math.*, 55(2):207–223, 2004.

[18] Jonathan Pila. Rational points on a subanalytic surface. *Ann. Inst. Fourier (Grenoble)*, 55(5):1501–1516, 2005.

[19] Andrea Surroca. Sur le nombre de points algébriques où une fonction analytique transcendante prend des valeurs algébriques. *C. R. Math. Acad. Sci. Paris*, 334(9):721–725, 2002.

[20] Andrea Surroca. Valeurs algébriques de fonctions transcendantes. *Int. Math. Res. Not., Art. ID 16834, 31pp*, 2006.

[21] E. C. Titchmarsh. *The theory of functions, second edition*. Clarendon, Oxford, 1939.

[22] Michel Waldschmidt. Transcendence measures for exponentials and logarithms. *J. Austral. Math. Soc. Ser. A*, 25(4):445–465, 1978.

[23] A. J. Wilkie. Model completeness results for expansions of the ordered field of real numbers by restricted Pfaffian functions and the exponential function. *J. Amer. Math. Soc.*, 9(4):1051–1094, 1996.

9

Ax-Schanuel and o-minimality

Jacob Tsimerman

1 Interpreting Ax-Schanuel Geometrically

The goal of this note is to give a geometric interpretation of what's commonly known as the Ax-Schanuel theorem due to Ax[1], and to give a model-theoretical proof of it. In this section, we recall the theorem in a few different forms; nothing in this section is original work. To start, let's recall the

Theorem 1.1 (Ax-Schanuel) *Let $f_1,\ldots,f_n \in \mathbb{C}[[t_1,\ldots,t_m]]$ be power series that are $\mathbb{Q}$-linearly independent modulo $\mathbb{C}$. Then we have the following inequality:*

$$\dim_{\mathbb{C}} \mathbb{C}(f_1,\ldots,f_n,e(f_1),\ldots,e(f_n)) \geq n + rank\left(\frac{\partial f_i}{\partial t_j}\right)_{\substack{1\leq i\leq n \\ 1\leq j\leq m}}$$

where $e(x) = e^{2\pi i x}$ and $\dim_K L$ is the transcendence degree of L over K.

To see the geometric implication of this theorem, let's restrict to the case where the power series f_i are convergent in some open neighborhood $B \subset \mathbb{C}^m$. Note that by the Seidenberg embedding theorem[1], it is sufficient to look at this case. Define the uniformizing map

$$\pi_n : \mathbb{C}^n \to (\mathbb{C}^\times)^n, \pi_n(x_1,\ldots,x_n) = (e(x_1),\ldots,e(x_n))$$

---

[1] Thanks to Martin Bays for pointing this out to the author. Thanks also to the referee, for pointing out that this is a somewhat standard argument, already used by Kirby [4] to give yet another proof of Ax-Schanuel.

O-Minimality and Diophantine Geometry, ed. G. O. Jones and A. J. Wilkie. Published by Cambridge University Press. © Cambridge University Press 2015.

and the subset $D_n \subset \mathbb{C}^n \times (\mathbb{C}^\times)^n$ to be the set

$$(\vec{x}, \vec{y}) \in D_n \iff \pi_n(\vec{x}) = \vec{y}.$$

Then we have a well defined map $\vec{f} : B \to D_n$ given by

$$\vec{f}(t_1, \ldots, t_m)$$
$$= (f_1(t_1, \ldots, t_m), \ldots, f_n(t_1, \ldots, t_m), e(f_1(t_1, \ldots, t_m)), \ldots, e(f_n(t_1, \ldots, t_m))).$$

Define $U \subset D_n$ to be the image of $\vec{f}$. U is then a complex analytic space, and it is easy to verify that

$$\dim_{\mathbb{C}}(U) = \mathrm{rank}\left(\frac{\partial f_i}{\partial t_j}\right)_{\substack{1 \le i \le n \\ 1 \le j \le m}}$$

and denoting the Zariski closure of U by U^{zar},

$$\dim_{\mathbb{C}}(U^{zar}) = \dim_{\mathbb{C}} \mathbb{C}(f_1, \ldots, f_n, e(f_1), \ldots, e(f_n)).$$

Moreover, denote by π_a and π_m the projections onto $\mathbb{C}^n$ and $(\mathbb{C}^\times)^n$ respectively[2]. Then the linear independence condition on the f_i is equivalent to saying that $\pi_a(U)$ does not lie in the translate of a proper $\mathbb{Q}$-linear subspace of $\mathbb{C}^n$, or that $\pi_m(U)$ does not lie in a coset of a proper subtorus. We can thus rephrase the Ax-Schanuel theorem geometrically as follows (this was observed by Ax in [2]):

Theorem 1.2 (Ax-Schanuel, V.2)   *Defining D_n, π_m as above, let $U \subset D_n$ be an irreducible complex analytic subspace such that $\pi_m(U)$ does not lie in a coset of a proper subtorus of $(\mathbb{C}^\times)^n$. Then*

$$\dim_{\mathbb{C}} U^{zar} \ge \dim_{\mathbb{C}} U + n$$

where U^{zar} denotes the Zariski closure of U in $\mathbb{C}^n \times (\mathbb{C}^\times)^n$.

It is more convenient to rephrase the above as a theorem about subvarieties of $\mathbb{C}^n \times (\mathbb{C}^\times)^n$. This is easy to do by starting with U^{zar} instead of U. The following rephrasing is then equivalent to the above:

Theorem 1.3 (Ax-Schanuel, V.3)   *Let $V \subset \mathbb{C}^n \times (\mathbb{C}^\times)^n$ be an irreducible subvariety, and let U be a connected, irreducible component of $V \cap D_n$. Assume that $\pi_m(U)$ is not contained in a coset of a proper subtorus of $(\mathbb{C}^\times)^n$. Then*

$$\dim_{\mathbb{C}} V \ge \dim_{\mathbb{C}} U + n.$$

---

[2] The reason for the notation is that the additive and multiplicative groups are denoted $\mathbf{G}_a$ and $\mathbf{G}_m$.

It is instructive to see how the above immediately implies the following corollary, dubbed 'Ax-Lindemann-Weirstrass' by J.Pila:

Corollary 1.4 *Suppose that $V_1 \subset \mathbb{C}^n$ and $V_2 \subset (\mathbb{C}^\times)^n$ are irreducible varieties with $\pi(V_1) \subset V_2$. Then there exists a coset of a torus S such that $S \subset V_2$ and $\pi(V_1) \subset S$.*

Proof We argue by induction on n, the case $n = 1$ being a trivial base case. Wlog we assume that V_2 is the Zariski closure of $\pi_n(V_1)$. If V_2 is contained in a coset of a proper subtorus $S \subset (\mathbb{C}^\times)^n$ we may identify S with $(\mathbb{C}^\times)^m$ for an integer $m < n$ and conclude by induction. Thus, we assume that V_2 is not contained in a coset of a proper subtorus.

Take $V = V_1 \times V_2$ in Theorem 1.3. It is clear that $V \cap D_n$ is the graph of V_1 under π_n. Thus, either $\pi_m(V)$ is contained in a coset of a proper subtorus S, or $\dim V \geq \dim V_1 + n$. In the first case it follows that $V_2 = \pi_n(V_1)^{zar} \subset S$ contradicting our assumption. In the latter case, we must have $V_2 = (\mathbb{C}^\times)^n$ and we may take $S = (\mathbb{C}^\times)^n$.   □

Acknowledgements. It is a pleasure to thank Jonathan Pila who introduced me to this circle of ideas and who carefully read over a previous version of the article, making suggestions that greatly improved the exposition. Moreover, Pila and Gareth Jones kindly alerted me to a problem in an earlier draft of the proof and suggested a fix. Finally, thanks to the referee for many useful comments on exposition.

2 An o-minimality proof of Ax-Schanuel

This entire section is devoted to a proof of Theorem 1.3 using the techniques of Pila-Zannier. We proceed by induction, the induction being lexicographic on the triple $(n, \dim V - \dim U, n - \dim U)$. To start with, we can assume that $U^{zar} = V$. As the case of U being a point is trivial, we assume U has positive dimension. By convention, definable always means definable in the o-minimal structure $\mathbb{R}_{an,exp}$. For more details on this field, see [3].

Definition 2.1 For an irreducible analytic set $X \subset \mathbb{C}^n \times (\mathbb{C}^\times)^n$, we define X^{Lin} to be the smallest affine linear subvariety containing $\pi_a(X)$.

Define

$$F = \{(z_1, \ldots, z_n, w_1, \ldots, w_n) \in \mathbb{C}^n \times (\mathbb{C}^\times)^n \mid 0 \leq Re(z_i) \leq 1\}$$

and note that $D_n \cap F$ is definable. Thus $U \cap F = V \cap D_n \cap F$ is definable. Moreover, for an analytic set $X \subset \mathbb{C}^n \times (\mathbb{C}^\times)^n$ and a linear subspace $L \subset \mathbb{C}^n$ we define $G_d(X, L)$ to be the set of points $x \in X$ around which X is regular of dimension d, and such that the irreducible component X_0 containing x satisfies X_0^{Lin} is a translate of L.

Let $I \subset \mathbb{R}^n$ be defined by

$$I = \{\ell \in \mathbb{R}^n \mid G_{\dim U}\left(((V + \ell) \cap (D_n \cap F)), U^{\text{Lin}}\right) \neq \emptyset\}$$

where addition is defined by acting on the first n co-ordinates of $\mathbb{C}^n \times (\mathbb{C}^\times)^n$. Then I is definable, and we're going to get somewhere by considering the intersection of I with $\mathbb{Z}^n$, the monodromy group of π_n.

Define $F_{\vec{m}} = F + \vec{m}$ and note that $\bigcup_{\vec{m} \in \mathbb{Z}^n} F_{\vec{m}} = \mathbb{C}^n \times (\mathbb{C}^\times)^n$. Moreover, if $U \cap F_{\vec{m}} \neq \emptyset$ then $-\vec{m} \in I$. This is because

$$(U \cap F_{\vec{m}}) - \vec{m} = (U - \vec{m}) \cap F \subset (V - \vec{m}) \cap D_n \cap F$$

where we have used the fact that $D_n + \vec{m} = D_n$. Assume first that $I \cap \mathbb{Z}^n$ is finite. In this case, it follows that U is a finite union of $U \cap F_{\vec{m}}$ and so is definable. Hence U is definable, closed and analytic in $\mathbb{C}^n \times (\mathbb{C}^\times)^n$, and so by [5, Theorems 4.5 and 5.3], U must be an algebraic variety. However, it is trivial to show that D_n contains no positive dimensional algebraic varieties (f and $e(f)$ can't both be algebraic functions for growth reasons, for example) which is a contradiction.

We thus conclude that $I \cap \mathbb{Z}^n$ is infinite. In particular, U intersects infinitely many $F_{\vec{m}}$. However, since U is connected the set of vectors $\vec{m}$ such that $U \cap F_{\vec{m}} \neq \emptyset$ must be a connected set in the graph G with vertex set $\mathbb{Z}^n$ and where the edges are given by connecting pairs of vertices all of whose co-ordinates are off by at most 1. It follows that $I \cap \mathbb{Z}^n$ has at least T integer points of height at most T. Applying the counting theorem of Pila-Wilkie ([6],Thm 1.9) we conclude that I contains a semi-algebraic curve $C_\mathbb{R}$, containing at least 1 smooth non-zero integer point $l \in C_\mathbb{R}(\mathbb{Z})$. We refer to the corresponding complex algebraic curve as C.

Next, consider the algebraic variety $V + C$. For each $c \in C_\mathbb{R}$ consider an irreducible component W_c of $(V + c) \cap (D_n \cap F)$ of dimension $\dim U$, such that W_c^{Lin} is a translate of U^{Lin}. If there are infinitely many such components as c varies, then there must be a component W of $(V + C) \cap D_n$ containing infinitely many such W_c. Hence W is of dimension at least $\dim U + 1$. Moreover, since $\pi_a(U)$ is not contained in a coset of a $\mathbb{Q}$-linear subspace it implies that U^{Lin} isn't and hence $\pi_a(W)$ isn't either. Thus we can replace V and U by $V + C$ and W and conclude by induction.

Otherwise, there must be only finitely many such W_c. Hence, there must be such a component $W = W_c$ appearing in infinitely many translates $V + c$, and thus by analyticity in all such translates by $c \in C$. If V is not invariant by translation under all elements of C, we replace (U, V) by $(W, \bigcap_{c \in C} V + c)$ and conclude by induction. Thus, we may assume that $V + C = V$.

In particular, V is invariant under l, hence also under the complex line generated by l by algebraicity. We make a linear change of co-ordinates with $\mathbb{Z}$-coefficients in $\mathbb{C}^n$ so that l is a multiple of $(1, 0, \ldots, 0)$ and the corresponding 'monomial' change of coordinates in $(\mathbb{C}^\times)^n$ so as to keep D_n invariant – note that this change of co-ordinates preserves all relevant dimension. We can thus write V as $V = \mathbb{C} \times V^0$ where

$$V^0 \subset \mathbb{C}^{n-1} \times (\mathbb{C}^\times)^{n-1} \times \mathbb{C}^\times.$$

The idea now is to apply induction on n.

So write $D_n = D_1 \times D_{n-1}$ and $U = \bigcup_{z \in D_1} \{z\} \times U_z$. For $z \in D_1$, let $V_z \subset \mathbb{C}^{n-1} \times (\mathbb{C}^\times)^{n-1}$ denote the fiber of V over z. Note that since U surjects under projection onto an open set of D_1, so must V and since V is algebraic it must be dominant onto $D_1^{zar} = \mathbb{C} \times \mathbb{C}^\times$. Thus, $\dim V = 2 + \dim V_z$ for a generic z. Now, we split into two cases:

- Suppose that the $\pi_a(U_z) \subset \mathbb{C}^{n-1}$ are not generically contained in a proper $\mathbb{Q}$-linear subspace. Then by induction, we have that for a generic z

$$\dim V_z \geq n - 1 + \dim U_z$$

  which yields our claim since $\dim V = \dim V_z + 2$ while $\dim U = \dim U_z + 1$.
- Else, since U is not contained in a proper $\mathbb{Q}$-linear subspace the U_z must vary with z. Let $U_0 \subset D_{n-1}$ be the projection of U and $V_0 \subset \mathbb{C}^{n-1} \times (\mathbb{C}^\times)^{n-1}$ be the projection of V. Note that since the U_z vary, we have that $\dim U = \dim U_0$. Then by induction, $\dim V_0 \geq \dim U_0 + (n - 1)$. This again yields our claim, since $\dim V \geq 1 + \dim V_0$.

References

[1] J. Ax, On Schanuel's Conjectures, *Annals of Mathematics*, Second Series, **93**, No. 2 (1971), 252–268.

[2] J. Ax, Some Topics in Differential Algebraic Geometry I: Analytic Subgroups of Algebraic Groups, *American Journal of Mathematics* **94**, No. 4 (1972), 1195–1204.

[3] L. van den Dries and C. Miller, On the real exponential field with restricted analytic functions, *Israel J. Math.* **85** (1994), 19–56.

[4] J. Kirby, A Schanuel condition for Weierstrass equations, *J. Symbolic Logic* **70**, Issue 2 (2005), 631–638.

[5] Y. Peterzil and S. Starchenko, Tame complex analysis and o-minimality, *Proceedings of the ICM, Hyderabad 2010*.

[6] J. Pila and A. J. Wilkie, The rational points of a definable set, *Duke Math. J.* **133** (2006), 591–616.

Printed in the United States
by Bookmasters

Printed in the United States
By Bookmasters